P9-DNE-713

Greenhouse Gases

Greenhouse Gases

Worldwide Impacts

Julie Kerr Casper, Ph.D.

GREENHOUSE GASES: Worldwide Impacts

Facts On File, Inc.
An imprint of Infobase Publishing
132 West 31st Street
New York NY 10001

Library of Congress Cataloging-in-Publication Data

Casper, Julie Kerr.
Greenhouse gases : worldwide impacts / Julie Kerr Casper.
p. cm.—(Global warming)
Includes bibliographical references and index.
ISBN 978-0-8160-7264-4
1. Greenhouse gases. 2. Global warming. I. Title.
TD885.5.G73C375 2009
363.738'74—dc22 2009004727

Facts On File books are available at special discounts when purchased in bulk quantities for businesses, associations, institutions, or sales promotions. Please call our Special Sales Department in New York at (212) 967-8800 or (800) 322-8755.

You can find Facts On File on the World Wide Web at http://www.factsonfile.com

Text design by Erik Lindstrom
Illustrations by Dale Williams
Composition by Hermitage Publishing Services
Cover printed by Bang Printing, Brainerd, MN
Book printed and bound by Bang Printing, Brainerd, MN
Date printed: February 2011
Printed in the United States of America

10 9 8 7 6 5 4 3 2

This book is printed on acid-free paper.

CONTENTS

Preface **viii**
Acknowledgments **xii**
Introduction **xiv**

1 The Greenhouse Effect and Global Warming **1**
Discovery of the Greenhouse Effect 1
Fourier, Arrhenius, and Callendar: Pioneers of Global Warming *3*
The Greenhouse Effect 5
Greenhouse Gases 18
Global Warming Potential 21
The Role of Ozone 25
Global Dimming 27

2 Carbon Sequestration **30**
The Earth's Carbon Cycle 30
Sinks and Sources 32
Types of Carbon Sequestration 39
The Carbon Conundrum *54*
Modeling Carbon Emissions in the United States 58

3 Agriculture and Greenhouse Gases **60**
The Effects of Global Warming on Agriculture 60
The Response of Agriculture to Climate Change 61
Acres of Opportunity—Greenhouse Gas Mitigation and Biofuels 74
Agriculture and Food Supply 80

4 **Deforestation and Greenhouse Gases** **87**
Forests' Role in Climate Change 87
Carbon Sinks 94
Impacts of Deforestation 98
Rain Forest Facts *102*
Climate Change Impacts 105
After Deforestation and the Future 109

5 **Anthropogenic Causes and Effects** **113**
Carbon Footprints 114
Calculating Your Carbon Footprint *114*
Land Use Change 119
Public Lands 123
Cities at Risk 128
Areas Most at Risk *133*
Climate Solutions for Cities *137*
Cultural Losses 138
Economic Impacts 139

6 **The Fate of Natural Refuges** **141**
Climate Change Impacts to Nature and Wildlife 141
The Endangered Species Act *145*
Working Together to Protect Wildlife *148*
The Timing of Seasonal Events 152
The Myths and Realities About the Endangered Species Act *154*
Wildlife in Peril *156*
Refuges at Risk 158
The Role of Refuges *159*
Laws in Force to Help Wildlife 166

7 **Global Warming around the World** **171**
Europe, Russia, and Asia 172
Eyewitness Accounts *177*
Africa and Oceania 179
The Arctic and Antarctica 181
South, Central, and North America 184
Global Warming Effects in the United States 186

8 **Mitigation and Adaptation to Climate Change** **204**
Impacts of Global Warming 204
Mitigation Strategies—Carbon Capture and Storage 208
Mitigation—One Step at a Time: Heading in the Right Direction *224*
Adaptation Strategies 224

Chronology **234**
Appendix: The Top 20 Carbon Dioxide Emitters **240**
Glossary **242**
Further Resources **248**
Index **261**

PREFACE

We do not inherit the Earth from our ancestors—
we borrow it from our children.

This ancient Native American proverb and what it implies resonates today as it has become increasingly obvious that people's actions and interactions with the environment affect not only living conditions now, but also those of many generations to follow. Humans must address the effect they have on the Earth's climate and how their choices today will have an impact on future generations.

Many years ago, Mark Twain joked that "Everyone talks about the weather, but no one does anything about it." That is not true anymore. Humans are changing the world's climate and with it the local, regional, and global weather. Scientists tell us that "climate is what we expect, and weather is what we get." Climate change occurs when that average weather shifts over the long term in a specific location, a region, or the entire planet.

Global warming and climate change are urgent topics. They are discussed on the news, in conversations, and are even the subjects of horror movies. How much is fact? What does global warming mean to individuals? What should it mean?

The readers of this multivolume set—most of whom are today's middle and high school students—will be tomorrow's leaders and scientists. Global warming and its threats are real. As scientists unlock the mysteries of the past and analyze today's activities, they warn that future generations may be in jeopardy. There is now overwhelming evidence that human activities are changing the world's climate. For thousands of years, the Earth's atmosphere has changed very little; but today, there are problems in keeping the balance. Greenhouse gases are being added to the atmosphere at an alarming rate. Since the Industrial Revolution (late 18th, early 19th centuries), human activities from transportation, agriculture, fossil fuels, waste disposal and treatment, deforestation, power stations, land use, biomass burning, and industrial processes, among other things, have added to the concentrations of greenhouse gases.

These activities are changing the atmosphere more rapidly than humans have ever experienced before. Some people think that warming the Earth's atmosphere by a few degrees is harmless and could have no effect on them; but global warming is more than just a warming—or cooling—trend. Global warming could have far-reaching and unpredictable environmental, social, and economic consequences. The following demonstrates what a few degrees' change in the temperature can do.

The Earth experienced an ice age 13,000 years ago. Global temperatures then warmed up 8.3°F (5°C) and melted the vast ice sheets that covered much of the North American continent. Scientists today predict that average temperatures could rise 11.7°F (7°C) during this century alone. What will happen to the remaining glaciers and ice caps?

If the temperatures rise as leading scientists have predicted, less freshwater will be available—and already one-third of the world's population (about 2 billion people) suffer from a shortage of water. Lack of water will keep farmers from growing food. It will also permanently destroy sensitive fish and wildlife habitat. As the ocean levels rise, coastal lands and islands will be flooded and destroyed. Heat waves could kill tens of thousands of people. With warmer temperatures, outbreaks of diseases will spread and intensify. Plant pollen mold spores in the air will increase, affecting those with allergies. An increase in severe weather could result in hurricanes similar or even stronger than Katrina in 2005, which destroyed large areas of the southeastern United States.

Higher temperatures will cause other areas to dry out and become tinder for larger and more devastating wildfires that threaten forests, wildlife, and homes. If drought destroys the rain forests, the Earth's delicate oxygen and carbon balances will be harmed, affecting the water, air, vegetation, and all life.

Although the United States has been one of the largest contributors to global warming, it ranks far below countries and regions—such as Canada, Australia, and western Europe—in taking steps to fix the damage that has been done. Global Warming is a multivolume set that explores the concept that each person is a member of a global family who shares responsibility for fixing this problem. In fact, the only way to fix it is to work together toward a common goal. This seven-volume set covers all of the important climatic issues that need to be addressed in order to understand the problem, allowing the reader to build a solid foundation of knowledge and to use the information to help solve the critical issues in effective ways. The set includes the following volumes:

Climate Systems
Global Warming Trends
Global Warming Cycles
Changing Ecosystems
Greenhouse Gases
Fossil Fuels and Pollution
Climate Management

These volumes explore a multitude of topics—how climates change, learning from past ice ages, natural factors that trigger global warming on Earth, whether the Earth can expect another ice age in the future, how the Earth's climate is changing now, emergency preparedness in severe weather, projections for the future, and why climate affects everything people do from growing food, to heating homes, to using the Earth's natural resources, to new scientific discoveries. They look at the impact that rising sea levels will have on islands and other areas worldwide, how individual ecosystems will be affected, what humans will lose if rain forests are destroyed, how industrialization and pollution puts peoples' lives at risk, and the benefits of developing environmentally friendly energy resources.

The set also examines the exciting technology of computer modeling and how it has unlocked mysteries about past climate change and global warming and how it can predict the local, regional, and global climates of the future—the very things leaders of tomorrow need to know *today.*

We will know only what we are taught;
We will be taught only what others deem is important to know;
And we will learn to value that which is important.
—Native American proverb

ACKNOWLEDGMENTS

Global warming may very well be one of the most important issues you will have to make a decision on in your lifetime. The decisions you make on energy sources and daily conservation practices will determine not only the quality of your life, but also those of your future descendants.

I cannot stress enough how important it is to gain a good understanding of global warming: what it is, why it is happening, how it can be slowed down, why everybody is contributing to the problem, and why *everybody* needs to be an active part of the solution.

I would sincerely like to thank several of the federal government agencies that research, educate, and actively take part in dealing with the global warming issue—in particular, the National Aeronautics and Space Administration (NASA), the National Oceanic and Atmospheric Administration (NOAA), the Environmental Protection Agency (EPA), and the U.S. Geological Survey (USGS) for providing an abundance of resources and outreach programs on this important subject. I would also like to acknowledge the monumental efforts of the Intergovernmental Panel on Climate Change (IPCC) and the unfailing efforts of more than 2,500 scientists worldwide dedicated for the past 20 years toward educating the public about the risks of climate change caused by human activity, as well as Dr. James E. Hansen at NASA's GISS, who has been involved in cutting-edge research for the past two decades and spent countless hours educating government leaders and academic institutions. I give special thanks to Al Gore and Arnold Schwarzenegger for their diligent efforts toward bringing the global warming issue so powerfully to the public's attention. I would also like to acknowledge and give thanks to the many wonderful universities across the United States, in England, Canada, and Australia, as well as private organizations, such as the World Wildlife Fund, Defenders of Wildlife, and the Union of Concerned Scientists, which diligently strive to educate others and help toward finding a solution to this very real problem.

I want to give a huge thanks to my agent, Jodie Rhodes, for her assistance, guidance, and efforts; and also to Frank K. Darmstadt, my editor, for all his hard work, dedication, support, and helpful advice and attention to detail. His efforts to bringing this project to life were invaluable. Thanks also to Alexandra Lo Re and the production and copyediting departments for their assistance and the outstanding quality of their work.

INTRODUCTION

As global warming increases in intensity, its effects are felt worldwide. While some geographical areas will be hit harder than others, and different areas will be affected in different ways, it is a global phenomenon and requires the effort of everyone on Earth to tackle and manage this growing problem—not only for us but also for our descendants.

While there are several factors that can contribute to global warming, such as natural changes in the Earth's inclination and revolution around the Sun, by far the biggest factor contributing to the global warming threat today is due to the emission of greenhouse gases—such as carbon dioxide (CO_2), methane, ozone, chlorofluorocarbons, water vapor, and nitrous oxide—added at alarming rates to the atmosphere by daily human activity. Every person on Earth has a "carbon footprint"—a measure of greenhouse gas contributed to the atmosphere on a daily basis. Some people's footprints are much higher than others; those who live in developed countries, such as the United States (which is the largest greenhouse gas emitter), emit much greater amounts than those in the undeveloped (unindustrialized) countries of the world. Various activities, such as agricultural and deforestation practices, also emit greenhouse gases.

Greenhouse Gases explores the very important role of greenhouse gases and their global impact on populations and ecosystems worldwide. Through gaining an understanding of the various sources of these gases, their interaction with the atmosphere, their effect on natural systems, and why controlling them is critical to the Earth's future climate, it can empower you—the reader—to take corrective action. One of the most critical aspects of the entire problem is public education. To date, there has been extensive research completed on global warming, and the consensus of the majority of the scientific community concurs that there is overwhelming evidence that global warming is happening right now and will continue to worsen if action is not taken immediately.

In order to make effective changes, the general population must first understand the issues and the consequences of their personal actions, and then be willing to make appropriate changes in their decisions and lifestyles. This volume will help you accomplish those goals.

Chapter 1 discusses the greenhouse effect and its role in global warming. You will learn about the various greenhouse gases—their similarities and differences—and why the concept of "global warming potential" is so important. It also discusses the role of ozone and another important newly discovered concept called "global dimming" and how it relates to global warming.

Next, chapter 2 discusses the process of carbon sequestration and why it is so important now and will become even more so in the future. It explores the concept of carbon sinks and sources and outlines the various types of carbon sequestration that are available. It also looks at how carbon emissions are modeled in the United States.

Chapter 3, explores the relationship of agriculture to greenhouse gases. A two-sided issue, you will learn how global warming affects agriculture as well as how agriculture responds to climate change. It also looks at how agriculture can be used as a tool to mitigate greenhouse gas emissions by both sequestration and biofuel production. Finally, it addresses the issue of adequate food supply in a warming world.

Chapter 4 discusses how deforestation impacts the greenhouse gas level in the atmosphere. It looks at forests' role in climate change and their role as global carbon sinks. The real impacts of deforestation are explored and what that means for climate change and the future given the current rate of tropical rain forest deforestation.

Chapter 5 deals with the pertinent issues of anthropogenic (human) causes and effects toward global warming. It discusses the concept of carbon footprints and shows you how to calculate your own. It looks at land use change and how it can either promote or curb greenhouse gas emissions. It also profiles which cities are most at risk, the role of public lands, which cultural sites around the world are being threatened and destroyed because of global warming, and what significant economic impacts are being experienced.

Chapter 6 examines the fate of natural refuges and how global warming is impacting nature and wildlife now and how scientists expect it to impact it in the future. It looks at the timing of seasonal events, such as migration, breeding, and food cycles. It also looks at threatened and endangered species and why global warming is adding more to that list. Finally, it looks at the various refuges that are currently at risk, their roles, and laws in force designed to help wildlife in peril due to climate change.

Chapter 7 takes you around the world for a look at the effect that the already changing climate is having in various geographical regions. It first looks at Europe, Russia, and Asia; then it explores the plight of Africa and Oceania. Next, it looks at the polar regions, then the Americas. Finally, it gives a close look at the impacts in the various regions of the United States and what types of climate scenarios are expected during the next century.

Finally, chapter 8 presents mitigation and adaptation principles applicable to climate change. It explores carbon capture and storage and various practical mitigation and adaptation strategies that can help curb the emission of greenhouse gases and slow global warming.

This book will give you a firm basis for an understanding of greenhouse gases, the critical role they play in global warming, and how you can help make a difference in reducing them. The statistical information used within the book is the most currently available data. This volume will also empower you with the knowledge to enlighten others to enable them to reduce their carbon footprint as well. After all, we are all in this together.

The Greenhouse Effect and Global Warming

Without the natural *greenhouse effect* on Earth, life as it exists today would not be possible. In order to understand the cause and effects of *global warming,* it is first necessary to understand the Earth's natural greenhouse effect and the roles of the various *greenhouse gases,* such as *carbon dioxide* (CO_2). This chapter looks at the difference between the Earth's natural greenhouse effect and the enhanced greenhouse effect and illustrates why and how humans are contributing to a global problem destined to ruin the health of the planet if not stopped. It also looks at the issue of global dimming, what it is, what causes it, its interaction with global warming, and what it means for the future.

DISCOVERY OF THE GREENHOUSE EFFECT

It was during the 19th century that scientists realized that gases—such as CO_2—found within the *atmosphere* cause a "greenhouse effect" that regulates the atmosphere's temperature. Ironically, the discovery of the greenhouse effect did not happen because scientists were trying to understand global warming; it happened because they were searching for the mechanism that triggered ice ages. In an effort to understand the connection between CO_2 and glacial periods, their interest was in studying the time intervals when concentrations of CO_2 were at their lowest, which correlated with the glacial periods in the Earth's past climate.

The beginning of the discovery process began with Joseph Fourier in the 1820s. During this time period, scientists were beginning to understand that the gases that composed the atmosphere may trap the heat received in the atmosphere from the Sun.

Also at this time, John Tyndall, a natural philosopher, was interested in finding out whether any gases in the atmosphere could actually trap heat rays. In 1859, through a series of lab analyses, he was able to identify several gases that were able to trap and hold heat. The most important of these gases were water vapor (H_2O) and CO_2.

Later, in 1896, Svante Arrhenius, a Swedish chemist/physicist, who was working with data on the prehistoric ice ages, was able to determine in his laboratory that by cutting the amount of CO_2 in the atmosphere by half, it could lower the temperature over Europe about 7–9°F (4–5°C)—roughly the equivalent of what would trigger another ice age.

In order for this to happen, however, the effect would have to be global. From this point, Arrhenius turned to Arvid Högbom, who added a modern twist to the analysis. He discovered that various human activities were adding CO_2 to the atmosphere at a rapid rate. At that time, he thought that the addition was not serious enough for alarm—it was not much different from other natural processes like erupting volcanoes. What he was concerned about, however, was that if the volumes continued being released into the atmosphere, it would not be long before they did start to negatively affect its quality. Arrhenius suggested that at the current rate of coal burning, the atmosphere could begin to start warming in a few centuries.

About that time, Thomas C. Chamberlin, an American geologist, became interested in atmospheric CO_2 levels, and the Swedish scientist Knut Ångström discovered that greenhouse gases do cause temperature to rise by retaining the heat instead of letting it escape to space. This added additional enlightenment to the beginning of the global warming theory.

FOURIER, ARRHENIUS, AND CALLENDAR: PIONEERS OF GLOBAL WARMING

Jean-Baptiste-Joseph Fourier, Svante Arrhenius, and G. S. Callendar are three scientists who are credited with the advancement of the science of global warming through their research and proactive leadership skills dedicated to dealing with the problem of global warming.

Jean-Baptiste-Joseph Fourier (March 21, 1768–May 16, 1830).
French physicist and mathematician.

Jean-Baptiste-Joseph Fourier began paving the way toward the understanding of the greenhouse effect. In 1824, his work led him to believe that the gases in the atmosphere could actually increase the surface temperature of the Earth. He postulated that the properties of the Earth's atmosphere actually warm the entire planet. Once he proposed this concept, it spurred thought that the Earth actually had an "energy balance." In other words, the Earth, through various processes, was able to receive, generate, and transfer heat within its own system. When the Earth obtains energy from other sources or from within itself, it can transfer heat by conduction or convection. When the Earth obtains energy from outside sources, such as the Sun, that can cause the temperature to rise. In response, when the Earth heats up, the heat energy is changed to "infrared" *radiation* (longer wavelengths), and heat is reradiated outward and lost.

There is a balance between heat gain and heat loss. Fourier correctly calculated that the amount of infrared radiation increases with temperature. He also identified the fact that the Earth receives energy from the Sun, in the form of solar radiation. This works effectively because most of the atmosphere is transparent to the Sun's insolation (incoming solar radiation), allowing it to enter the atmosphere.

Svante Arrhenius (February 19, 1859–October 2, 1927).
Swedish chemist.

Svante Arrhenius, from Vik, Sweden, contributed to scientific knowledge of the greenhouse effect and is credited with its discovery. Interested in developing a theory to explain ice ages, he originally speculated that it was changing levels of CO_2 in the atmosphere that could significantly alter the Earth's climate via the greenhouse effect. Many of Arrhenius's ideas were based upon the work and breakthroughs of Fourier. Arrhenius theorized in 1908 that the human emission of CO_2 would be strong enough to

(continues)

(continued)

prevent the world from entering another ice age. He was the first person to predict that emissions of CO_2 from the burning of *fossil fuels* and other processes that use combustion machines would lead to global warming.

Still focused on ice ages, Arrhenius, however, believed that a warmer world would be better overall than a cooling one. He also made some predictions about the Earth's atmosphere based on varying the amounts of CO_2. He calculated that if the current CO_2 level was lowered by half, it would cause a cooling of 5.7–8.3°F (4–5°C). If, on the other hand, the CO_2 was doubled, it would cause the temperature to rise 7–11°F (5–6°C). What is especially interesting about his calculations is that based on the report last issued from the Intergovernmental Panel of Climate Change (*IPCC*) released in 2007, they project a temperature rise of 3.3–7.5°F (2–4.5°C). The similarity between Arrhenius's and the IPCC's results are nearly the same. The only difference is that Arrhenius predicted it would take the CO_2 about 3,000 years to double and cause this level of temperature rise; the IPCC has projected it will take only about 100 years.

Guy Stewart Callendar, (February 1898–October 1964).
Engineer, inventor, and amateur meteorologist.

G. S. Callendar, an amateur meteorologist, compiled a reliable global dataset of surface temperatures. He documented an upward trend in temperatures between 1900 and 1940. He was also able to correlate them in a cause-and-effect scenario to the retreat of the world's *glaciers* and the rising CO_2 in the atmosphere as compared to the same areas before the *Industrial Revolution*. He also studied the infrared absorption bands and concluded that the trend toward higher temperatures was significant.

He was able to determine that it was the increase in fossil fuels that had caused the atmospheric CO_2 to rise about 10 percent from the 1800s and that the increase in radiation had contributed to the rising temperatures. He was also credited with the development of the "Callendar Effect," which is defined as climate change brought about by *anthropogenic* increases in the concentration of atmospheric CO_2.

The invaluable contribution of these scientists has allowed for the advancements seen today in the understanding of global warming and the effects it has had, is having now, and will have in the future on the environment.

THE GREENHOUSE EFFECT

The Earth needs the natural greenhouse effect. It is the process in which the *emission* of infrared radiation by the atmosphere warms the planet's surface. The atmosphere naturally acts as an insulating blanket, which is able to trap enough solar energy to keep the global average temperature in a comfortable range in which to support life. This insulating blanket is actually a collection of several atmospheric gases, some of them in such small amounts that they are referred to as trace gases.

The framework in which this system works is often referred to as the greenhouse effect because this global system of insulation is similar to that which occurs in a greenhouse nursery for plants. The gases are relatively transparent to incoming visible light from the Sun, yet opaque to the energy radiated from the Earth.

These gases are the reason why the Earth's atmosphere does not scorch during the day and freeze at night. Instead, the atmosphere contains molecules that absorb and reradiate the heat in all directions, which reduces the heat lost back to space. It is the greenhouse gas molecules that keep the Earth's temperature ranges within comfortable limits. Without the natural greenhouse effect, life would not be possible on Earth. In fact, without the greenhouse effect to regulate the atmospheric temperature, the Sun's heat would escape and the average temperature would drop from 57°F to –2.2°F (14°C to –18°C); a temperature much too cold to support the diversity of life that exists today on the planet.

Solar Radiation

The spectrum of the Sun's solar radiation is similar to that of a black body with a temperature of 5,800 K. Roughly half of it lies within the range of the visible shortwave portion of the *electromagnetic spectrum.* The other half resides mostly in the near infrared portion, just beyond the visible wavelengths. A small amount lies in the ultraviolet range.

Incoming solar radiation, also referred to as insolation, is a measure of the solar radiation energy that is received on a specific surface area for a specific amount of time. It can be expressed as average irradiance in watts per square meter (W/m^2) or kilowatt-hours per square meter per day ($kW{\cdot}h/m^2{\cdot}day$). The surface in this case would be the surface of the Earth. When the insolation strikes a surface, some of it will be absorbed, causing radiant heating of the object, and the rest will be reflected. Each surface reacts differently, depending on its orientation, texture, reflectivity, and other physical properties. The amount of radiation that gets either absorbed or reflected depends on the object's *albedo.* An object with a high albedo will reflect most of the isolation reaching it; an object with a low albedo will absorb most of the isolation reaching it. Surfaces with a high albedo usually remain cooler, such as an ice sheet. Objects with a low albedo usually heat up faster because they absorb more energy (in the form of heat), such as asphalt roads in urban areas.

The direction, or orientation, of an object also influences how the insolation reacts with it. Insolation is greatest when a surface faces the Sun. Surfaces at a large angle from the Sun result in less insolation. The larger the angle, the less insolation reaches it without being reflected off. This is why the polar areas are colder than the equatorial areas.

Several interactions can occur to insolation once it enters the Earth's atmosphere. If it does not have to interact with any particulates, gases, or other components in the atmosphere, then much of the energy will make it to the Earth's surface as "direct insolation." If, however, there are a lot of particulates in the atmosphere, which interfere with the incoming radiation, it will cause the solar radiation to be scattered and reflected in a process called "diffuse insolation."

The incoming solar radiation travels as wavelengths at the speed of light. Over a year, the average solar radiation arriving at the top of the Earth's atmosphere is about 1,366 W/m^2; and the radiant power is distributed across the entire electromagnetic spectrum. Although the Sun's radiant power is distributed across the entire electromagnetic spectrum, most of it is centered in the visible portion of the spectrum. When the Sun's rays enter the atmosphere, they are attenuated—or weakened—so that they are roughly 1,000 W/m^2 for a surface that is perpendicular to the Sun's rays at sea level on a clear day.

Radiation Transmission

The Earth reflects about 30 percent of the incoming solar radiation. The other 70 percent is absorbed and warms the atmosphere, the land, and the oceans. In order to maintain an energy balance, so the Earth does not get too hot or cold, the amount of incoming energy must roughly equal the amount of outgoing energy. The insolation occurs in the visible portion of the spectrum. The radiation that is emitted from objects that have been heated on Earth, however, peaks in the infrared portion of the spectrum in the form of heat. The Earth's energy balance is maintained when the incoming solar radiation is roughly balanced by the outgoing energy being radiated by the Earth back out into space as infrared radiation.

The point of heating and radiation varies between the insolation and the outgoing radiation. The visible solar radiation heats the Earth's surface, not the atmosphere. The infrared radiation the Earth gives off in response, however, is emitted to space from the Earth's upper atmosphere, not the Earth's surface. The greenhouse gases in the atmosphere absorb any infrared radiation that does originate from the Earth's surface, where it then stays.

There are many paths that the energy from the Sun to the Earth's atmosphere and surface can take. Energy interacts with the Earth's surface, clouds, and atmospheric gases as it undergoes various routes: It can be directly reflected off the Earth's surface, absorbed and then reradiated, and then reabsorbed by the Earth's atmosphere. The illustration on page 8 shows the unique way the Earth's atmosphere is able to capture and redirect the energy.

The atmosphere allows roughly 70 percent of the Sun's radiation to reach the Earth, reflecting the remainder back to outer space. As depicted by the illustration, the majority of insolation is in the short and medium wavelength region of the spectrum. The highest energy, short wavelengths, such as the gamma rays, X-rays, and ultraviolet (UV) light, is absorbed by the mid to high levels of the atmosphere. This is desirable because if these wavelengths traveled to the surface they could cause harm to life on Earth. For example, exposure to UV light can lead to cancer. The medium wavelengths—referred to as visible light—travel to the Earth's surface. These waves can be absorbed or reflected at the Earth's surface, as well as by the CO_2 and water vapor in the atmosphere.

When wavelengths from the electromagnetic spectrum reach the Earth, several things can happen. They can be reflected; the Earth's

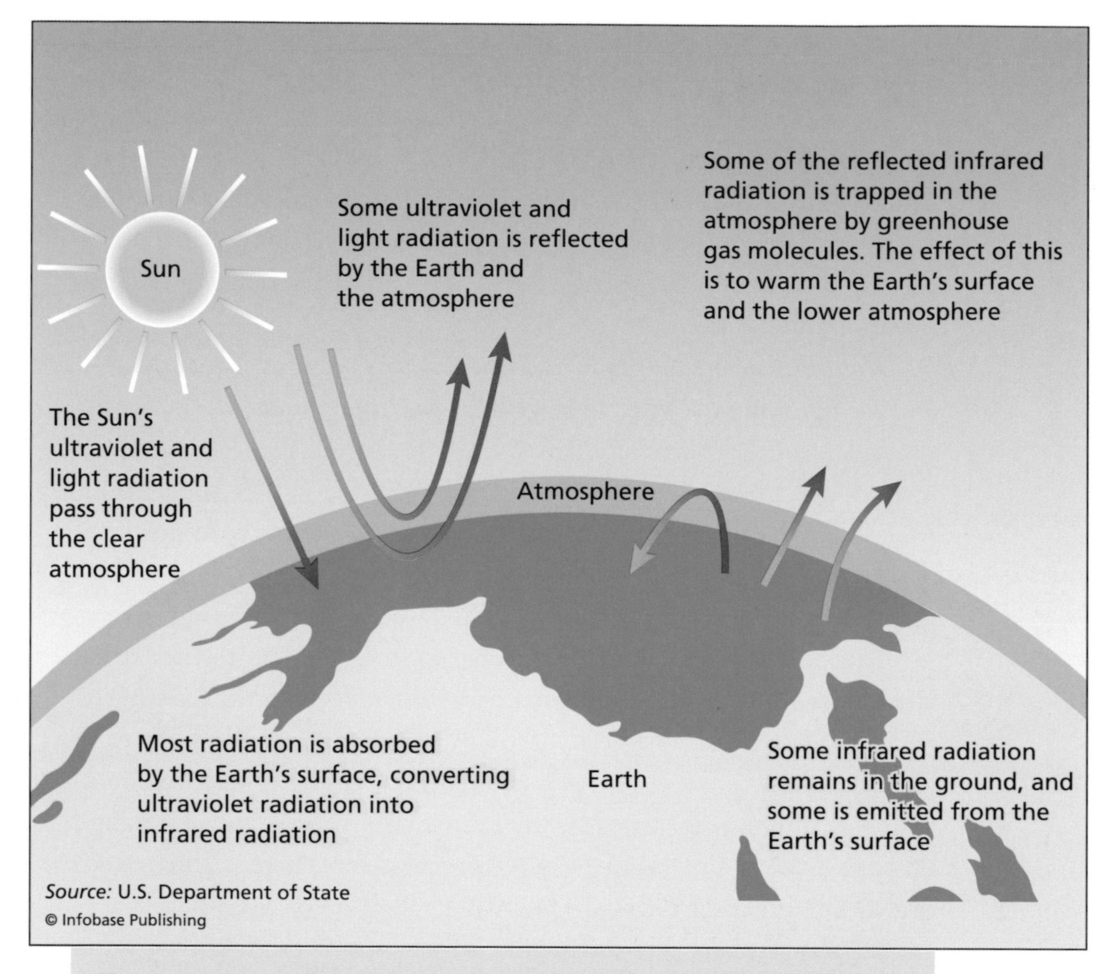

The greenhouse effect refers to the interaction between the Earth's atmosphere and surface to absorb, transfer, and emit energy as heat, cycling it through the atmosphere and back to the surface. The natural greenhouse effect is necessary for life as it exists on Earth today. *(Source: GRID)*

surface, the ocean, the clouds, or the atmosphere can absorb them; and they can be scattered. Roughly one-third of the incoming visible light is directly reflected back into space without ever reaching the Earth's surface. Of that which does enter the Earth's atmosphere, nearly one-quarter of that is reflected by clouds and particulates (tiny suspended particles) in the atmosphere. Of the remaining visible light that does reach the Earth's surface, approximately one-tenth is reflected upward by snow and ice because these surfaces have an extremely high albedo (high reflectivity).

Of the two-thirds of the visible light that is transmitted successfully through the Earth's atmosphere, half is absorbed by the land and water on the Earth's surface, and one-fifth is absorbed by the clouds and atmosphere. When electromagnetic energy is absorbed at the Earth's surface, it heats it up and then reradiates it back as longwave infrared (IR) radiation. The temperature is also much cooler: 210–310 K. Infrared radiation is what keeps the air warm above the Earth's surface.

As the infrared radiation is emitted upward, it does not just escape to space. Water vapor and other gases in the atmosphere absorb it and

The Earth's natural greenhouse effect is often compared to a nursery greenhouse because of its ability to alter the local environment, keeping heat confined in a specific environment. *(Nature's Images)*

trap it in the atmosphere. This trapped heat then radiates in all directions: upward, down toward the Earth's surface again, and sideward. This trapped heat energy is what constitutes the greenhouse effect. Because the heat is being "held in" by the atmosphere, it functions in a similar manner to the concept of a nursery greenhouse, warming the Earth's environment.

The Natural and Enhanced Greenhouse Effect

In this "greenhouse environment," it is the combination of water vapor and trace gases that are responsible for trapping the heat radiated from the Sun. The natural amount keeps the Earth habitable. Without this trapped warmth, the Earth would have an average temperature of –0.4°F (–18°C). If this were the case, the Earth would look much different—there would most likely be very little, if any, liquid water available. The entire Earth would have an *ecosystem* similar to that of the harshest areas in Antarctica.

The Earth's natural greenhouse effect is critical for the survival and diversity of life. Since the Industrial Revolution (over the past 250 years or so), the natural greenhouse effect has been augmented by human interference. CO_2, one of the atmosphere's principal greenhouse gases, has been altered to such an extent by human activity that the natural greenhouse effect is no longer in balance. The Earth's energy balance now must contend with what is referred to as the "enhanced greenhouse effect" or "anthropogenic greenhouse warming." CO_2 is being added in voluminous amounts as a result of human activity; *deforestation*; agricultural practices; the burning of fossil fuels for transportation; urban development; heating and cooling homes; and industrial processes.

CO_2 in the atmosphere has increased 31 percent since 1895. Concentration of other greenhouse gases, such as *methane* and *nitrous oxides* (also related to human activity) have increased 151 percent and 17 percent, respectively.

This enhanced greenhouse effect began in the age of the Industrial Revolution of the 1700s. At that time, CO_2 content in the atmosphere was 280 *parts per million (ppm)*. By 1958, it had increased to 315 ppm; by 2004, 378 ppm; by 2005, 379 ppm; and by 2007, 383 ppm. According to James E. Hansen, a world-renowned expert on global warming and *climate* change in New York at NASA's Goddard Institute for Space Studies (GISS), "Climate is nearing dangerous tipping points. We have already gone too far. We must draw down atmospheric CO_2 to preserve the planet we know. A level of no more than 350 ppm is still feasible, with the help of reforestation and improved agricultural practices, but just barely—time is running out."

Although not the only greenhouse gas, CO_2 does seem to be the one most focused on because it is one of the most important and prevalent. It makes up about 25 percent of the natural greenhouse effect. It is so prominent because there are several natural processes that put it in the atmosphere, such as the following:

- Forest fires: When trees are burned, the CO_2 stored within them is released into the atmosphere.
- Volcanic eruptions: CO_2 is one of the gases released in abundance by volcanoes.
- Oceans: Oceans both absorb and release enormous amounts of CO_2. Historically, they have served as a major storage facility of CO_2 put into the atmosphere from human sources. Recently, however, scientists at NASA and NOAA have determined that the oceans have become nearly saturated and are approaching their limits as a carbon store.
- Trees, plants, grasses, and other vegetation serve as significant stores of *carbon*. When they die and decompose, half of their stored carbon is released into the atmosphere in the form of CO_2.
- When vegetation dies, the other half of their stored carbon is absorbed by the soil. Over time, some of this carbon is released into the atmosphere as CO_2.
- Any life-form that consumes plants that contain carbon also emit CO_2 into the atmosphere through breathing. This includes animals, insects, and even humans.

With all this CO_2 finding its way into the atmosphere, fortunately there are some processes that help keep it in check. Plants and trees remove CO_2 from the atmosphere through *photosynthesis* (turning the Sun's energy into food). The oceans absorb large amounts of CO_2; and phytoplankton (tiny organisms that float in the ocean) also take in CO_2 through photosynthesis.

Ideally, keeping CO_2 in balance is desirable, but throughout the Earth's history, this has not always been possible. When CO_2 levels have dropped, the Earth has consistently experienced an ice age. Since the last ice age, the CO_2 levels have remained fairly constant, however, until the Industrial Revolution, when billions of tons of extra CO_2 began to be added to the Earth's atmosphere.

The turning point was the use of fossil fuels—coal, oil, and gas. These fuels are made from the carbon of plants and animals that decomposed millions of years ago and were buried deep under the Earth's surface, subjected to enormous amounts of heat and pressure. When this carbon is converted into usable energy forms and is burned as a fuel, the carbon combines with oxygen and releases enormous amounts of CO_2 into the atmosphere.

According to NASA/GISS, CO_2 levels have not been as high as they are today over the last 400,000 years. The enhanced greenhouse effect has not come as a surprise to climate scientists, other experts, and decision makers, however. In addition to being addressed by Fourier, Arrhenius, and Callendar, another climate giant, Charles David Keeling of the Scripps Institution of Oceanography, made enormous strides in establishing the rising levels of CO_2 in the Earth's atmosphere and the subsequent issue of the enhanced greenhouse effect and global warming.

Keeling set up a CO_2 monitoring station at Mauna Loa, Hawaii, in 1957. Beginning in 1958, Keeling made continuous measurements of CO_2, which still continue today. He chose the remote Mauna Loa site so that the proximity of large cities and industrial areas would not compromise the atmospheric readings he was collecting. Air samples at Mauna Loa are collected continuously from air intakes at the top of four 23-foot (7 m) towers and one 89-foot (27 m) tower. Four air samples are collected each hour to determine the CO_2 level.

The measurements he gathered show a steady increase in mean atmospheric concentration from 315 parts per million by volume (ppmv) in 1958 to more than 383 ppmv by 2007. The increase is considered to be largely due to the enhanced greenhouse effect and the

The world's first functioning CO_2 monitoring station, initiated by Charles Keeling, is atop Mauna Loa, the highest mountain on Hawaii. Here scientists are able to monitor atmospheric CO_2 levels. This is a good place to measure CO_2 levels because it is far away from major cities, where large amounts of air pollution could interfere with the results. This facility has been instrumental in determining that CO_2 levels are rising; the resultant graph from the continual monitoring since 1957 is one of the most well-documented pieces of evidence supporting global warming. *(NOAA)*

burning of fossil fuels and deforestation. This same upward trend also matches the paleoclimatic data shown in ice cores obtained from Antarctica and Greenland. The ice cores also show that the CO_2 *concentration* remained at approximately 280 ppmv for several thousand years, but began to rise sharply at the beginning of the 1800s.

The Keeling Curve also shows a cyclic variation of about 5 ppmv each year. This is what gives it the "sawtooth" appearance. It represents the seasonal uptake of CO_2 by the Earth's vegetation (principally in the Northern Hemisphere because that is where most of the Earth's

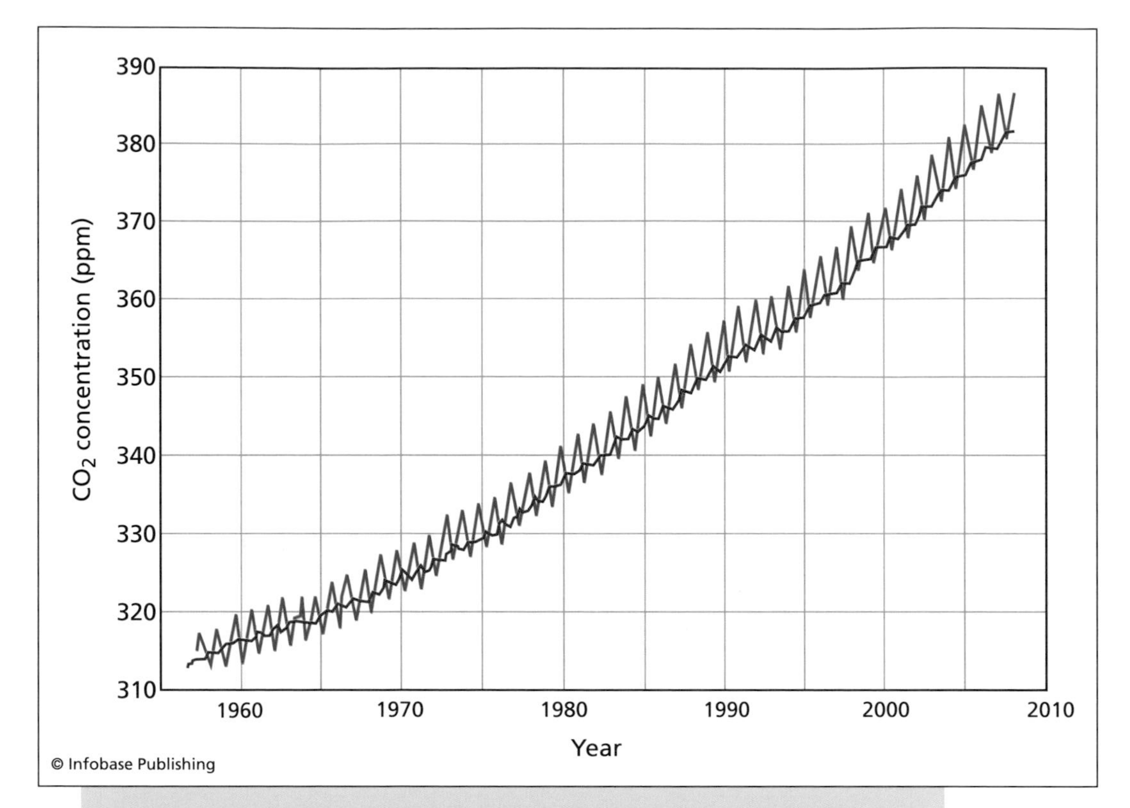

Collected since 1958, the Keeling Curve has been instrumental in providing convincing evidence that CO_2 levels are rising worldwide. The blue line depicts the readings collected from the Mauna Loa observation station, and the red line depicts data collected from the South Pole. Without this continuous data, it would be much more difficult to determine the existence of global warming.

vegetation is located). The CO_2 lowers in the spring because new plant growth takes CO_2 out of the atmosphere through photosynthesis; and it rises in autumn as plants and leaves die off and decay, releasing CO_2 gas back into the atmosphere.

Mauna Loa is known today as the premier long-term atmospheric monitoring facility on Earth. It is also considered one of the world's most favorable locations for measuring undisturbed air because vegetation and human influences are minimal and any influence from volcanic vents have been calculated and excluded from the measurement. The methods and equipment used to create the continuous monitoring record have been the same for the past 50 years, so the monitoring program is highly controlled and consistent.

According to the Carbon Dioxide Research Group of the Scripps Institution of Oceanography, "Because of the favorable site location, continuous monitoring, and careful selection and scrutiny of the data, the Mauna Loa record is considered to be a precise record and a reliable indicator of the regional trend in the concentrations of atmospheric CO_2 in the middle layers of the troposphere."

Charles Keeling was a true pioneer in establishing the existence of global warming. His work was instrumental in showing scientists worldwide that the CO_2 level in the atmosphere was indeed continually rising. This was the first strongly compelling evidence that global warming not only existed, but was accelerating. Because of the worthwhile merits and significance of his work, NOAA began monitoring CO_2 levels worldwide in the 1970s. There are currently about 100 sites continuously monitored today because of what was started on Mauna Loa.

The Keeling Curve has now celebrated its 50th anniversary of continuous collection, representing the world's longest continuous record of atmospheric CO_2 levels. This single curve is what provided the first compelling evidence that atmospheric CO_2 levels have been rising since the middle of the 20th century. Ralph Keeling, Charles Keeling's son, says, "The Keeling Curve has become an icon of the human imprint on the planet and a continuing resource for the study of the changing global carbon cycle. Mauna Loa provides a valuable lesson on the importance of continuous Earth observations in a time of accelerating global change."

The Intergovernmental Panel on Climate Change (IPCC) has projected that with continued global warming, temperatures will rise an average of 2.3–9.7°F (1.4–5.8°C) in the next 100 years. Polar region temperatures, however, are expected to rise more than the average. Winter temperatures over land could rise as much as 4.2–23.3°F (2.5–14°C) above current normal temperatures. This could happen because of a change in surface albedo. Currently, with much of the polar areas being covered by snow and ice, they have a high albedo, so much of the insolation is reflected back to space. As temperatures warm, however, the snow and ice will melt, exposing darker surfaces (land and water), which will naturally absorb more heat, which will melt more ice and uncover more dark surface area. This could continue in a cycle until all the snow and ice are melted, greatly changing the heat balance in the polar regions, causing a rapid warming.

Radiative Forcing

When discussing the enhanced greenhouse effect and global warming, climate scientists often refer to a concept called "radiative forcing." This is a measure of the influence that an independent factor (ice albedo, *aerosols, land use,* carbon dioxide) has in altering the balance of incoming and outgoing energy in the Earth-atmosphere system. It is also used as an index of the influence a factor has as a potential climate change mechanism. Radiative *forcings* can be positive or negative. A positive forcing (corresponding to more incoming energy) warms the climate system. A negative forcing (corresponding to more outgoing energy) cools the climate system.

As shown in the illustration, examples of positive forcings (those which warm the climate) include greenhouse gases, tropospheric *ozone,* water vapor, solar irradiance, and anthropogenic effects. Those that have a negative rating (they cool the climate) include stratospheric ozone and certain types of land use. *Climatologists* often use radiative forcing data to compare various cause-and-effect scenarios in climate systems where some type of change (warmer or cooler) has taken place. Radiative forcings and their effects on climate are also built into *climate models* in order to enable climatologists to determine the effects from multiple input factors on climate change. The greenhouse gases that remain in the atmosphere the longest will have the greatest impact on global warming, making it imperative that climatologists not only be able to identify them, but also understand them and how they react with the atmosphere.

The Earth's Energy Balance

Even if greenhouse gas emissions were stabilized at today's rates, global warming would not stop. According to scientists at NASA, CO_2 levels are rising because there is more CO_2 being emitted into the atmosphere than natural processes such as photosynthesis and absorption into the

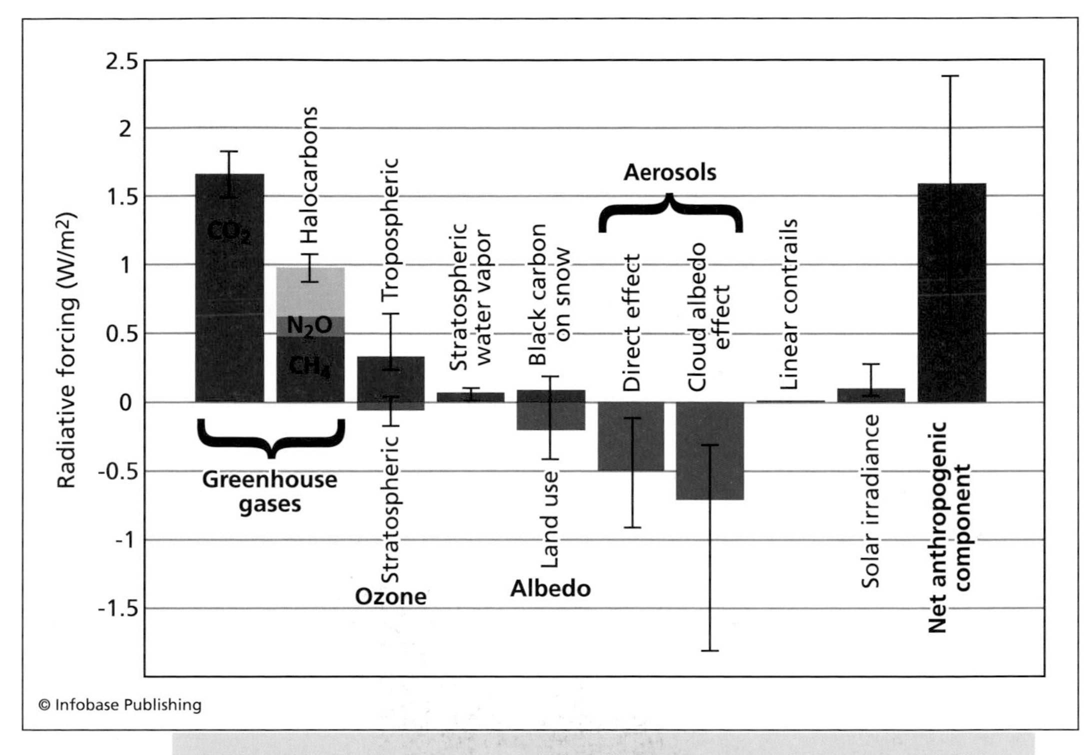

Positive radiative forcings warm the climate; negative forcings cool the climate. One of the largest positive forcings today is the human-caused component leading to global warming.

oceans can presently remove. There is no longer a carbon balance; there is a net gain that continues to increase each year. As the CO_2 levels increase, the temperatures continue to rise.

Because the Earth's carbon system is already so out of balance, in order to stop global warming, emissions must actually be reduced over the coming years, not just stabilized. Even with the best of intentions, the Earth's temperature would not react immediately—there is a delayed reaction. This occurs because there is already an abundance of excess energy stored in the world's oceans. This "lag time" responding to the reaction is referred to as "thermal inertia." Because of *thermal* inertia, scientists at NASA have determined that the 1–1.5°F (0.6–0.9°C) of global warming that has already occurred this past century is not the full amount of warming the environment will eventually reach from the greenhouse gases that have already been emitted into the atmosphere. Even in the extreme case that all greenhouse emissions stopped immediately, the Earth's average surface temperature would continue to climb 1°F (0.6°C) or more over the next several decades before temperatures leveled off.

The half-life of CO_2 is about 100 years. Therefore, most of the CO_2 being released today will still be in the atmosphere in 2109. Although there are natural factors that control the CO_2 levels and climate, NASA supports the notion that human influence is changing the climate on Earth—many agree today's documented global warming trend is at least partly anthropogenic in origin. The IPCC states: "The balance of evidence suggests that there is a discernible human influence on global climate."

According to James E. Hansen at NASA/GISS, this lag time is a key reason why it is risky to hold off any longer trying to control greenhouse gas emissions. The longer society waits to take positive action to stop global warming, the more severe and long-lasting the consequences will become. He stresses that the time to act is now, not some time in the future.

If global warming is allowed to continue, there will be many negative effects, such as the disappearance of ecosystems, leading to the extinction of many species; there will be a greater number of heat-related deaths; there will be a greater spread of infectious diseases, such as encephalitis, malaria, and dengue fever through the proliferation of disease-carrying mosquitoes; malnutrition and starvation as a result of droughts; loss of coastal areas due to sea level rise from melting ice caps and thermal expansion of the oceans; unpredictable agricultural production; threatened food supply and contaminated water. Impacts will be health-related, social, political, ecological, environmental, and economic.

GREENHOUSE GASES

It is the existence of trace gases in the atmosphere that act like the glass in a greenhouse. The trace gases serve to trap the heat energy from the Sun close to Earth. Most greenhouse gases occur naturally. Greenhouse gases are cycled through the global biogeochemical system. It is the greenhouse gases added by human activity that are trapping too much heat today and causing the atmosphere to overheat. There are several different types of greenhouse gases, some existing in greater quantities than others. They include water vapor, CO_2, methane, nitrous oxide, and halocarbons. Greenhouse gases capture 70 to 85 percent of the energy in upgoing thermal radiation emitted from the Earth's surface.

Water vapor is the most common greenhouse gas—it accounts for roughly 65 percent of the natural greenhouse effect. Water exists in many places on the Earth's surface—lakes, rivers, and oceans. When water heats up, it evaporates into vapor and rises from the Earth's surface into the atmosphere. It forms clouds and can act as a insulating blanket to help keep the Earth warm, or it can reflect and scatter incoming sunlight. This is why cloudy nights are warmer than clear nights. As water vapor condenses and cools, it then comes back to the Earth as snow or rain and continues on its way through the water cycle.

CO_2 is generated from several sources. The second most prevalent greenhouse gas, it comprises about 25 percent of the natural greenhouse effect. Humans and animals exhale CO_2; vegetation releases CO_2 when it dies and decomposes; burning trees in a forest fire or burning during deforestation release it; burning fossil fuels (such as exhaust from cars and industrial processes) are all common sources of CO_2.

Methane is a colorless, odorless, flammable gas formed when plants decay in an environment of very little air. It is the third-most common greenhouse gas and is created when organic matter decomposes without the presence of oxygen—a process called anaerobic decomposition. One of the most common sources is from "ruminants"—grazing animals that have multiple stomachs in which to digest their food. These include cattle, sheep, goats, camels, bison, and musk ox. In their digestive system, their large fore-stomach hosts tiny microbes that break down their food. This process creates methane gas, which is released as flatulence. Livestock also emit methane when they belch. In fact, in one day a single cow can emit one-half pound of methane into the air. Each day 1.3 billion cattle burp methane several times per minute. Humans also produce methane.

Methane is also a by-product of natural gas and decomposing organic matter, such as food and vegetation. Also present in wetlands, it is commonly referred to as "swamp gas." Since 1750, methane has doubled its concentration in the atmosphere, and it is projected to double again by 2050. According to Nick Hopwood and Jordan Cohen at the University of Michigan, every year 350 to 500 million tons of methane are added to the atmosphere through various activities, such as the raising of livestock, coal mining, drilling for oil and natural gas, garbage sitting in landfills, and rice cultivation. Rice cultivation is a huge global business. In the past 50 years, rice farmland has doubled in area. Rice is a major food staple; it currently feeds one-third of the world's population. Because rice is grown in waterlogged soils, like swamps, they release methane as a by-product.

Nitrous oxide (N_2O), another greenhouse gas, is released from manure and chemical fertilizers that are nitrogen-based. As the fertilizer breaks down, N_2O is released into the atmosphere. Nitrous oxide is also contained in soil by bacteria. When farmers plow the soil and disturb the surface layer, N_2O is released into the atmosphere. It is also released from catalytic converters in cars and also from the ocean. According to Hopwood and Cohen, nitrous oxide has risen more than 15 percent since 1750. Each year 7–13 million tons (6–12 million metric tons) is added to the atmosphere principally through the use of nitrogen-based fertilizers, the disposal of human and animal waste in sewage treatment plants, automobile exhaust, and other sources that have not been identified yet. The use of nitrogen-based fertilizers has doubled in the last 15 years. Although good for the productivity of crops, they break down in the soil and release N_2O into the atmosphere.

Halocarbons include the fluorocarbons, methylhalides, carbon tetrachloride (CCl_4), carbon tetrafluoride (CF_4), and halons. They are all considered to be powerful greenhouse gases because they strongly absorb terrestrial infrared radiation and stay in the atmosphere for many decades.

Fluorocarbons are a group of synthetic organic compounds that contain fluorine and carbon. A common compound is chlorofluorocarbon (CFC). This class contains chlorine atoms and have been used in industry as refrigerants, cleaning solvents, and propellants in spray cans. These fluorocarbons are harmful to the atmosphere, however, because they deplete the ozone layer; their use has been banned in most areas of the world, including the United States.

Hydrofluorocarbons (HFCs) contain fluorine and do not damage the ozone layer. Fluorocarbon polymers are chemically inert and electrically insulating. They are used in place of CFCs because they do not harm or break down ozone molecules, but they do trap heat in the atmosphere. HFCs are used in air conditioners and refrigerators. The best way to keep HFCs out of the atmosphere is to recycle the coolant from the equipment they are used in.

Fluorocarbons have several practical uses. They are used in anesthetics in surgery, as coolants in refrigerators, as industrial solvents, as lubricants, water repellents, stain repellents, and chemical reagents. They are used to manufacture fishing line and are contained in products such as Gore-Tex and Teflon.

GLOBAL WARMING POTENTIAL

Although there are several types of greenhouse gases, they are not the same—they all have different properties associated with them. For example, they have different lifetimes in which they reside in the atmosphere, and they also differ in the amount of heat that they can trap.

Many of the greenhouse gases are extremely potent—some can continue to reside in the atmosphere for thousands of years after they have been emitted. For instance, according to the EPA, they can be 140 to 23,900 times more potent than CO_2 in terms of their ability to trap and hold heat in the atmosphere over a 100-year period. It is important to note that these gases and their effects will continue to increase in the atmosphere as long as they continue to be emitted and remain there. Even though these gases represent a very small proportion of the atmosphere—less than two percent of the total—because of their enormous heat-holding potential, they are a significant component to the atmosphere and represent a serious problem to global warming.

The EPA has identified three major groups of high *Global Warming Potential* (GWP) gases: (1) hydrofluorocarbons (HFCs), (2) perfluorocarbons (PFCs), and (3) sulfur hexafluoride (SF_6). These represent the most potent greenhouse gases; and the PFCs and SF_6 also have extremely long atmospheric lifetimes—up to 23,900 years. Because their lifetime is so incredibly long, for practical management purposes, once they are emitted into the atmosphere, they are considered to be there permanently. According to the EPA, once present in the atmosphere, it results in "an essentially irreversible accumulation."

Hydrofluorocarbons are man-made chemicals; most of them were developed as replacements for the prior used ozone-depleting substances that were common in industrial, commercial, and consumer products. The GWP index for HFCs ranges from 140 to 11,700, depending on which one is used. Their lifetime in the atmosphere ranges from one to 260 years; the most commonly used ones have a lifetime of about 15 years and are used in automobile air-conditioning and refrigeration.

Perfluorocarbons generally originate from the production of aluminum and semiconductors. PFCs have very stable molecular structures and usually do not get broken down in the lower atmosphere. When they reach the mesosphere 37 miles (60 km) above the Earth's surface, high-energy ultraviolet electromagnetic energy destroy them, but it is a very slow process, which enables them to accumulate in the atmosphere for several thousand years (up to 50,000).

Sulfur hexafluoride has a GWP of 23,900, making it the most potent greenhouse gas. It is used in insulation, electric power transmission equipment, in the magnesium industry, in semiconductor manufacturing to create circuitry patterns on silicon wafers, and also as a tracer gas for leak detection. Its accumulation in the atmosphere shows the global average concentration has increased by 7 percent per year during the 1980s and 1990s—according to the IPCC from less than one part per trillion (ppt) in 1980 to almost four ppt in the late 1990s.

In order to understand the potential impact from specific greenhouse gases, they are rated as to their global warming potential. The GWP of a greenhouse gas is the ratio of global warming—or radiative forcing—from one unit mass of a greenhouse gas to that of one unit mass of CO_2 over a period of time, making the GWP a measure of the "potential for global warming per unit mass relative to CO_2." In other words, greenhouse gases are rated on how potent they are compared to CO_2.

GWPs take into account the absorption strength of a molecule and its atmospheric lifetime. Therefore, if methane has a GWP of 23 and carbon has a GWP of 1 (the standard), this means that methane is 23 times more powerful than CO_2 as a greenhouse gas. The IPCC has published reference values for GWPs of several greenhouse gases. Reference standards are also issued and supported by the United Nations Framework Convention on Climate Change (UNFCCC), as shown in the table.

Global Warming Potential of Greenhouse Gases		
GREENHOUSE GAS	LIFETIME IN THE ATMOSPHERE	GWP OVER 100 YEARS (COMPARED TO CO_2)
Carbon Dioxide	50–200 years	1
Methane	12 years	23
Nitrous Oxide	120 years	296
CFC 115	550 years	7,000
HFC-23	264 years	11,700
HFC-32	5.6 years	650
HFC-41	3.7 years	150
HFC–43–10mee	17.1 years	1,300
HFC-125	32.6 years	2,800
HFC-134	10.6 years	1,000
HFC-134a	14.6 years	1,300
HFC-152a	1.5 years	140
HFC-143	3.8 years	300
HFC-143a	48.3 years	3,800
HFC-227ea	36.5 years	2,900
HFC-236fa	209 years	6,300
HFC-245ca	6.6 years	560
Sulphur hexafluoride	3,200 years	23,900
Perfluoromethane	50,000 years	6,500
Perfluoroethane	10,000 years	9,200
Perfluoropropane	2,600 years	7,000

(continues)

(continued)

Global Warming Potential of Greenhouse Gases		
GREENHOUSE GAS	LIFETIME IN THE ATMOSPHERE	GWP OVER 100 YEARS (COMPARED TO CO_2)
Perfluorobutane	2,600 years	7,000
Perfluorocyclobutane	3,200 years	8,700
Perfluoropentane	4,100 years	7,500
Perfluorohexane	3,200 years	7,400
Source: UNFCCC		

The higher the GWP value, the larger the infrared absorption and the longer the atmospheric lifetime. Based on this table, even small amounts of SF_6 and HFC-23 can contribute a significant amount to global warming.

In response to global warming, the U.S. Environmental Protection Agency is working to reduce the emission of high GWP gases because of their extreme potency and long atmospheric lifetimes. High GWP gases are emitted from several different sources. Major emission sources of these today are from industries such as electric power generation, magnesium production, semiconductor manufacturing, and aluminum production.

In electric power generation, SF_6 is used in circuit breakers, gas-insulated substations, and switchgear. During magnesium metal production and casting, SF_6 serves as a protective cover gas during the processing. It improves safety and metal quality by preventing the oxidation and potential burning of molten magnesium in the presence of air. It replaced sulfur dioxide (SO_2), which was more environmentally toxic. The semiconductor industry uses many high GWP gases in plasma etching and in cleaning chemical vapor deposition tool chambers. They are used to create circuitry patterns. During primary aluminum production, GWP gases are emitted as by-products of the smelting process.

The best solution found to date to solve the negative impact to the environment and combat global warming is by the EPA working with private industry in business partnerships that involve developing and implementing new processes that are environmentally friendly. In addition, the EPA is also working to limit high GWP gases through mandatory recycling programs and restrictions.

If a greenhouse gas can remain in the atmosphere for several hundred years, even though it may be in a small amount, it can do a substantial amount of damage. Some of the greenhouse effect today is due to greenhouse gases put in the atmosphere decades ago. Even trace amounts can add up significantly.

THE ROLE OF OZONE

According to scientists at NASA, global warming and ozone depletion are two separate issues, but they do share some related aspects. Whereas global warming and the greenhouse effect refer to the lower portions of the atmosphere (the *troposphere*), enhanced by the continual addition of heat-trapping gases from human activity, ozone depletion refers to the upper portion of the atmosphere (the stratosphere). Disturbance of the ozone layer presents a serious health concern due to a lack of ability to block the Earth's surface from harmful ultraviolet radiation from the Sun. Even though they are technically separate issues, there is a connection between the two.

Scientists have found holes in the ozone layer high above the Earth. Ozone holes are not just empty spaces (lacking ozone) in the sky. Ozone holes are much like the worn-out places in a blanket: There are still threads covering the worn-out area, but the fabric can be so thin it can be seen through. The 1990 Clean Air Act provided provisions for fixing the holes, but repairs will take time.

Ozone in the stratosphere—the layer of the atmosphere 9 to 31 miles (14 to 50 km) above the Earth—serves as a protective shield, filtering out harmful solar radiation, including ultraviolet B. Exposure to ultraviolet B has been linked to the development of cataracts (eye damage) and skin cancer.

In the mid-1970s, scientists suggested that chlorofluorocarbons (CFCs) could destroy stratospheric ozone (the same CFC mentioned previously as a greenhouse gas with a lifetime in the atmosphere of 550 years). CFCs were widely used then as aerosol propellants in consumer products such as hairsprays and deodorants and for many uses in industry. Because of concern about the possible effects of CFCs on the ozone layer, in 1978 the U.S. government banned CFCs as propellants in aerosol cans.

Since the aerosol ban, scientists have been monitoring and measuring the ozone layer. A few years ago, an ozone hole was found above Antarctica, including the area of the South Pole. This hole, which has been appearing each year during the Antarctic winter (the Northern Hemisphere's summer), is bigger than the continental United States in size. More recently, ozone thinning has been found in the stratosphere above the northern half of the United States; the hole extends over Canada and up into the Arctic regions (the area of the North Pole). Initially, the hole was detected only in winter and spring, but more recently has continued into summer. Between 1978 and 1991, there was a 4 to 5 percent loss of ozone in the stratosphere over the United States, which represents a significant loss. Ozone holes have also been found recently over northern Europe. There is also evidence that holes in the ozone layer damage plant life. Scientists are looking into possible harm to agriculture, as well.

Because of this threat to the ozone layer, 93 nations—including the United States—have jointly agreed to reduce their production and use of ozone-destroying chemicals. When researchers discovered that the ozone layer was thinning more quickly than first thought, this agreement was revised to speed up the phase-out of ozone-destroying chemicals.

Unfortunately, this problem does not have an easy fix. It will be a long time before the ozone layer will be repaired. Because of the ozone-destroying chemicals already in the stratosphere and those that will arrive within the next few years, ozone destruction will likely continue for another 20 years. As substitutes are developed for ozone-destroying substances, before the chemicals can be produced and sold, the U.S. Environmental Protection Agency (EPA) must determine that the replacements will be safe for health and the environment.

There are connections between global warming and the ozone layer. The CFCs that destroy the ozone layer are also powerful greenhouse gases. Scientists at NASA believe there is a slight connection between the two. CFCs only destroy ozone at very cold temperatures—below −112°F (−80°C). Greenhouse gases absorb heat at a low altitude, which warm the surface but cool the stratosphere. They believe that the cooler the stratosphere, the more rapidly ozone may be destroyed. This cycle could make the ozone hole bigger.

The ozone layer does trap heat. Reducing ozone-depleting gases (such as fluorocarbons) is important in preventing the ozone layer's destruction, just as efforts to limit all types of emissions to limit global warming will also positively affect the ozone layer.

GLOBAL DIMMING

Global dimming is an important concept that accompanies global warming. Little understood until recently, it is a phenomenon that plays a role in the overall environmental health of the Earth and has specific implications for global warming.

Although global dimming has existed since as long as pollution has existed, it gained particular attention during the terrorist attacks on the United States on September 11, 2001, when all commercial airline traffic was grounded for three days. During this brief time frame, scientists discovered that the atmospheric temperature rose 1.7°F (1°C) during that short interval. As they studied the possibilities, they realized it was because there were no contrails in the atmosphere left by commercial airlines.

The trails of water vapor were reflecting some of the incoming solar radiation, preventing it from reaching the Earth's surface. This discovery led climate scientists at NASA and NOAA to understand that activities that reflect insolation may act counter to global warming. In other words, particulates in the atmosphere—pollutants (sulphur dioxide, soot, ash) and water vapor—while unhealthy in themselves, causing health conditions such as respiratory illnesses, actually serve to counteract some of the ill effects of global warming by reflecting heat away from the Earth.

Although this may seem like a possible solution to counter global warming, it is not a viable option because global dimming has its own associated negative impacts. Climatologists at NASA believe it has led to

Global dimming caused by airline contrails was discovered in 2001 to partially offset the effects of global warming. *(Nature's Images)*

decreased rainfall in the Sahel in Northern Africa, where it has caused severe drought and famine, causing heat-related deaths of more than a million people in Africa and has harmed 50 million people with hunger and starvation.

One of the harsh realities about global dimming, and one that has scientists worried, is that as cities deal with cleaning up pollution and reducing the levels of particulates in the atmosphere, it will increase the effects of global warming. Many believe that global dimming has been counteracting the negative effects of global warming. If only global dimming issues are addressed, then the negative effects of global warming will increase even more.

One worrisome aspect this has in the scientific community is that the counteracting effect of global dimming has not been adequately calculated into present-day climate models. Therefore, as the atmosphere gets cleaned up, it may trigger an even sharper rise in global warming, harming ecosystems even faster than anticipated. Experts believe this might negatively affect billions of people.

In fact, based on the *Horizon* documentary aired by the British Broadcasting Corporation in January 2005, scientists believe that dealing with global dimming without also dealing with the effects of global warming would cause several serious impacts with irreversible damage in only about two decades. It would increase the melting of the world's ice sheets; dry the *tropical* rain forests, increasing fire danger and thereby release more CO_2 into the atmosphere; kill vegetation including crops and devastate the agriculture industry; degrade soil; and release enormous amounts of methane hydrate from the ocean floor, which is about eight times more powerful as a greenhouse gas than CO_2. If global warming is not dealt with at the same time, it is projected that temperatures could rise 17°F (10°C) over the next 100 years—which is double what most models are predicting today.

To deal successfully with the issues of the various interactions of the Sun's radiation with the Earth and its atmosphere—the greenhouse gases, their interactions, and life cycles, as well as the interactions between global warming and global dimming—requires the joint research and attention of scientists, planners, managers, politicians, and the general public alike. An understanding of the current issues involving greenhouse gases and an applied knowledge of what will happen to the environment if they are not controlled and managed properly is half the battle facing the world population today. Once the issues are understood and accepted, moving on to the next step—informed positive action—can successfully occur.

Carbon Sequestration

CO_2 is the most prevalent greenhouse gas in the atmosphere. Because of that, it is important to understand carbon: what it is used for, where it comes from, where it goes, and how it interacts with other elements. As it relates to global warming, the concept of carbon sequestration—or carbon storage—is especially important. If carbon can be stored, it is removed from the environment, and, therefore, removed from being a potential source of greenhouse gas. This chapter looks at the role of carbon and the cycle it goes through—where it comes from and where it goes. It looks at which areas on Earth are sources of carbon and which are *carbon sinks* (reservoirs), able to store excess carbon from the Earth's atmosphere. Carbon sinks play a direct role in global warming—the more that can be sequestered, the less available to contribute to global warming.

THE EARTH'S CARBON CYCLE

The *carbon cycle* is important on Earth because carbon is the basic structural material for all cell life. Everything on Earth that is organic contains carbon. The carbon cycle involves the movement of carbon between the atmosphere, the oceans, the land, and living organisms.

Similar to the hydrologic cycle, where water can cycle long-term for thousands of years through groundwater or the oceans, or via short-term paths through streams and condensation, carbon also cycles on both a long- and short-term time frame. The long-term cycle—also referred to as the "geological scale"—cycles over time periods that encompass millions of years. The short-term cycle—referred to as "biological/physical scale"—operates in time frames ranging from several days to thousands of years.

Carbon cycles globally as it moves from one location to another. When carbon is being stored in a reservoir—either short- or long-term—it is referred to as a sink. There are multiple combinations of carbon "sources" and "sinks" on Earth that carbon cycles through—or exchanges between—over time. At any one time on Earth, there are an abundance of physical, chemical, biological, and geological processes occurring that involve carbon directly and influence its exchange and movement. Because the oceans cover three-fourths of the Earth's surface, they contain the largest active source of carbon near the Earth's surface. The carbon that is stored in the deeper levels of the ocean, however, does not get readily exchanged; it remains stored in the depths long term. There are four principal repositories, or reservoirs, of carbon on Earth, and they are exchanged through specific routes. The major conduits of carbon movement are as follows:

- oceans,
- terrestrial biosphere,
- sediments,
- atmosphere.

Carbon in the oceans is found in both living and nonliving marine biota as well as dissolved inorganic carbon. The terrestrial biosphere is the carbon found on the surface of the Earth's land. This includes living and nonliving organic material, the carbon that resides in soil, and carbon in freshwater systems. Sediments encompass geologic formations and also include fossil fuel deposits such as coal, oil, and gas. Because fossil fuels originated from buried organic matter (prehistoric plants and animals) formed under high pressure, they are high in carbon content. The atmosphere acts as a pathway for carbon to combine with oxygen as a gas and move through the cycle as CO_2. In addition to CO_2, there is also carbon in methane and chlorofluorocarbons, as explained in chapter 1. Therefore, carbon in the form of CO_2, carbonates, and organic compounds are cycled between the main reservoirs of the Earth's biogeochemical system.

Over various short- and long-term time scales, the global carbon cycle exists in dynamic equilibrium. Currently, the Earth's atmosphere stores around 827 billion tons (750 billion metric tons) of carbon in the form of CO_2. Of this, the deep oceans store more than 42,000 tons (38,000 metric tons), and the surface of the ocean contains about 1,102 tons (1,000 metric tons).

Over the past 250 years or so, the Earth's carbon cycle has become unbalanced. Because human sources of CO_2 have increased (such as from fossil fuels), the system is no longer in equilibrium. The atmospheric portion of the cycle has increased in far greater proportions than in other parts of the cycle. According to the IPCC, since pre-industrial times, atmospheric CO_2 concentrations have increased roughly 27 percent from about 280 ppm to 383 ppm. Restoring the equilibrium is not easily done. Because carbon is exchanged on both short- and long-term scales, fairly rapid transfers occur between the atmosphere and the biosphere.

SINKS AND SOURCES

In terms of the global warming issue, storing CO_2 in reservoirs—or sinks—is desirable in order to keep the CO_2 out of the atmosphere so that it does not contribute to ever-increasing atmospheric temperatures. There are several mechanisms whereby carbon can be put into storage.

During the process of photosynthesis, plants convert CO_2 from the atmosphere into carbohydrates and release oxygen during the process. Trees in forests over a period of years are able to store large amounts of carbon, which is why promoting the regrowth of old forests and encouraging the growth of new forests are positive methods of combating global warming. Vegetation sequesters 600 billion tons (544 billion metric tons) of carbon each year. In fact, the regrowth of forests in the Northern Hemisphere is the most significant anthropogenic sink of atmospheric CO_2. When forests are destroyed during the process of deforestation and the land is cleared for other uses such as grazing, it removes a valuable carbon sink from the ecosystem. In areas where the forests are cleared through burning, this process serves to release the CO_2 back into the atmosphere.

In the oceans, as living organisms with shells die, pieces of the shells break apart and fall toward the bottom of the ocean, slowly accumulating as a sediment layer on the bottom. Small amounts of plankton settle through the deep water and come to rest on the floors of the ocean basins. Even though this source does not represent huge sources of carbon, over thousands or millions of years it does add up, becoming another useful long-term carbon store.

Long-term exchanges involve storing anthropogenic-induced CO_2 into the deep ocean, where it will stay in a sink for a significant amount of time. The world's oceans are currently holding huge amounts of CO_2, helping to minimize the effects of global warming. As the oceans become saturated, however, they will not be able to store as much CO_2 as they have previously. Currently, about half of all the anthropogenic carbon emissions are absorbed by the ocean and the land.

At the ocean's surface, the water becomes cooler as it flows toward the polar areas. As water cools, the CO_2 contained within it becomes more soluble. The temperature of the ocean plays a critical part in how the seawater behaves. The currents in the ocean transport enormous amounts of heat. As water is moved from the equator to the poles, the seawater becomes cooler, which makes the CO_2 more soluble. Therefore, in the cold water areas of the world's oceans—such as the polar regions—more CO_2 is absorbed from the atmosphere by the cold ocean water. Conversely, in the warmer ocean waters—such as the equatorial region—the ocean releases CO_2 to the atmosphere.

As global warming continues and the ocean water temperatures also continue to rise, the oceans will not be able to store as much CO_2 as they have been—the warmer the water, the less CO_2 it will hold; more CO_2 than expected will remain in the atmosphere to heat the air. This exacerbating problem will only increase the effects of global warming, creating a vicious cycle. The phytoplankton that currently exists in the oceans are also important organisms for taking in and storing carbon.

According to a report in *National Geographic News* in June 2001, a team of 23 scientists led by Stephen Pacala, a professor of ecology and evolutionary biology at Princeton University, ran a study to determine how much carbon is stored in sinks in the United States. There are two different methods that can be used to determine an amount—one based on an atmospheric reading, the other on a ground reading. Intending to collect data as thoroughly as possible, the researchers gathered data from carbon absorbed by houses, soils, landfills, and silt in reservoirs.

According to Pacala, "We found out that the land sink was bigger than had been reported by other analyses—about twice as big."

The scientists determined that the forests and other land areas in the United States absorb from one-third to two-thirds of a billion tons (30 to 60 billion metric tons) of carbon every year. In contrast, the United States emits about 1.4 billion tons (1.3 billion metric tons). According to the researchers, the amount of carbon being stored in sinks in the United States is currently so large because many of the areas that were once "clear-cut" and logged to make way for agriculture and urbanization in the past century are now regrowing new vegetation, which is also serving as a sink.

Pacala also points out that part of this sequestration is the land reabsorbing large quantities of carbon that were released during intense farming and logging activities in the past. "When we chopped down the forests, we released carbon trapped in the trees into the atmosphere. When we plowed up the prairies, we released carbon from the grasslands and soils into the atmosphere. Now the ecosystem is taking some of that back. But, the sink effect will steadily decrease and eventually disappear as U.S. ecosystems complete their recovery from past land use. The carbon sinks are going to decrease at the same time as our fossil fuels emissions increase. Thus, the greenhouse problem is going to get worse faster than we expected."

A sink can also become a source if the situation changes. Carbon can also enter the cycle from several different sources. One of the most common is through the respiration process of plants and animals. Once a plant or animal dies, carbon also enters the system. Through the decaying process, the bacteria and fungi break down the carbon compounds. If oxygen is present, the carbon is converted into CO_2. If oxygen is not present, then methane is produced instead. Either way, both are greenhouse gases. A major source into the carbon cycle is through the combustion of *biomass*. This oxidizes the carbon, which produces CO_2. The most prevalent source of this is through the use of fossil fuels—coal, petroleum products, and natural gas. Fossil fuels are large deposits of biomass that have been commonly preserved inside geologic rock formations in the Earth's crust for millions of years. While they are preserved in rock formations, there is no release of carbon to the atmosphere. However, as soon as they are processed into fuel products and used for transportation and industrial processes, millions of tons are released into the atmosphere. According to the Environmental Protection Agency (EPA), the burning of fossil fuels and deforestation releases nearly 8.8 billion tons (8 billion metric tons) of carbon into the atmosphere each year.

Another major source of CO_2 in the atmosphere is from forest fires. When trees burn, the stored carbon is released to the atmosphere. Other gases such as carbon monoxide and methane are also released. After a fire has occurred and deadfall is left behind, it will further decompose and release additional carbon into the atmosphere. One of the predictions with global warming is that as temperatures rise, certain areas will experience increased drought conditions. When this occurs, vegetated areas will become extremely dry; then, when lightning strikes, wildfires can easily start, destroying vegetation, increasing the concentration of CO_2 gases in the atmosphere and warming it further.

Another source of carbon occurs near urban settings. When cement is produced, one of its principal components—lime (produced from the heating of limestone)—produces CO_2 as a by-product. Therefore, in urban areas this is another factor to consider; areas of increasing urbanization are areas of increasing CO_2 concentrations.

Volcanic eruptions can be a significant source of CO_2 during a major event. Enormous amounts of gases can be released into the atmosphere, including greenhouse gases such as water vapor, CO_2, and sulfur dioxide. CO_2 can also be released from the ocean's surface into the atmosphere from bursting bubbles. Phytoplankton, little organisms

Forest fires can add significant amounts of greenhouse gases to the atmosphere, such as carbon dioxide, carbon monoxide, and methane. *(Bureau of Land Management)*

floating in the world's oceans, also use some of the CO_2 to make their food through photosynthesis.

The Earth's zones of *permafrost* are another potential source of CO_2. According to the Oak Ridge National Laboratory in Tennessee, if climate temperatures continue to rise and cause the permafrost soils in the boreal (northern) forests to thaw, the carbon released from them could dramatically increase atmospheric CO_2 levels and ultimately affect the global carbon balance.

According to Mike Goulden, a scientist working on a NASA project to determine the net exchange of carbon in a Canadian forest, "The soils of boreal forests store an enormous amount of carbon."

He says the amount of carbon stored in the soil is greater than both the carbon stored in the moss layer and wood of the boreal forests. Boreal forests cover about 10 percent of the Earth's land surface and contain 15 to 30 percent of the carbon stored in the terrestrial biosphere. Most of the carbon is located in deep layers 16–31 inches (40–80 cm) below the surface, where it remains frozen—the zone called permafrost.

The reason that the permafrost layer is so high in carbon is that the decomposing plant material accumulates on the forest floor and becomes buried before it decomposes. With the steady increase in atmospheric temperatures over recent years, some of the permafrost has been experiencing deep-soil warming and been melting in midsummer. This process releases CO_2 through increased respiration.

According to Goulden, "Assessing the carbon balance of forest soil is tricky, but if you've got several pieces of evidence all showing the same thing, then you can really start to believe that the site is losing soil carbon. Our findings on the importance of soil thaw and the general importance of the soil carbon balance is already influencing the development of models for carbon exchange."

What Goulden is hoping to accomplish in the future is to find a way to use *satellite* data to improve carbon exchange models.

There are several human-related sources of CO_2. By far, the largest human source is from the combustion of fossil fuels such as coal, oil, and gas. This occurs in electricity generation, industrial facilities, from automobiles in the transportation sector, and from heating homes and commercial buildings. According to the EPA, in 2006, petroleum supplied the largest share of domestic energy demands in the United States. This equates to 47 percent of the total fossil fuel–based energy consumption during that year. Coal and natural gas followed, at 27 and 26 percent, respectively, all heavy producers of CO_2. Electricity generation is the single largest source of CO_2 emissions in the United States, according to the EPA. This is one reason why the EPA began their ENERGY STAR® program, geared toward the production and use of energy efficient appliances.

Industrial processes that add significant amounts of CO_2 to the atmosphere include manufacturing, construction, and mining activities. The EPA has identified six industries in particular that use the majority of energy sources, thereby contributing the largest share of CO_2: chemical production, petroleum refining, primary metal production, mineral production, paper production, and the food processing industry.

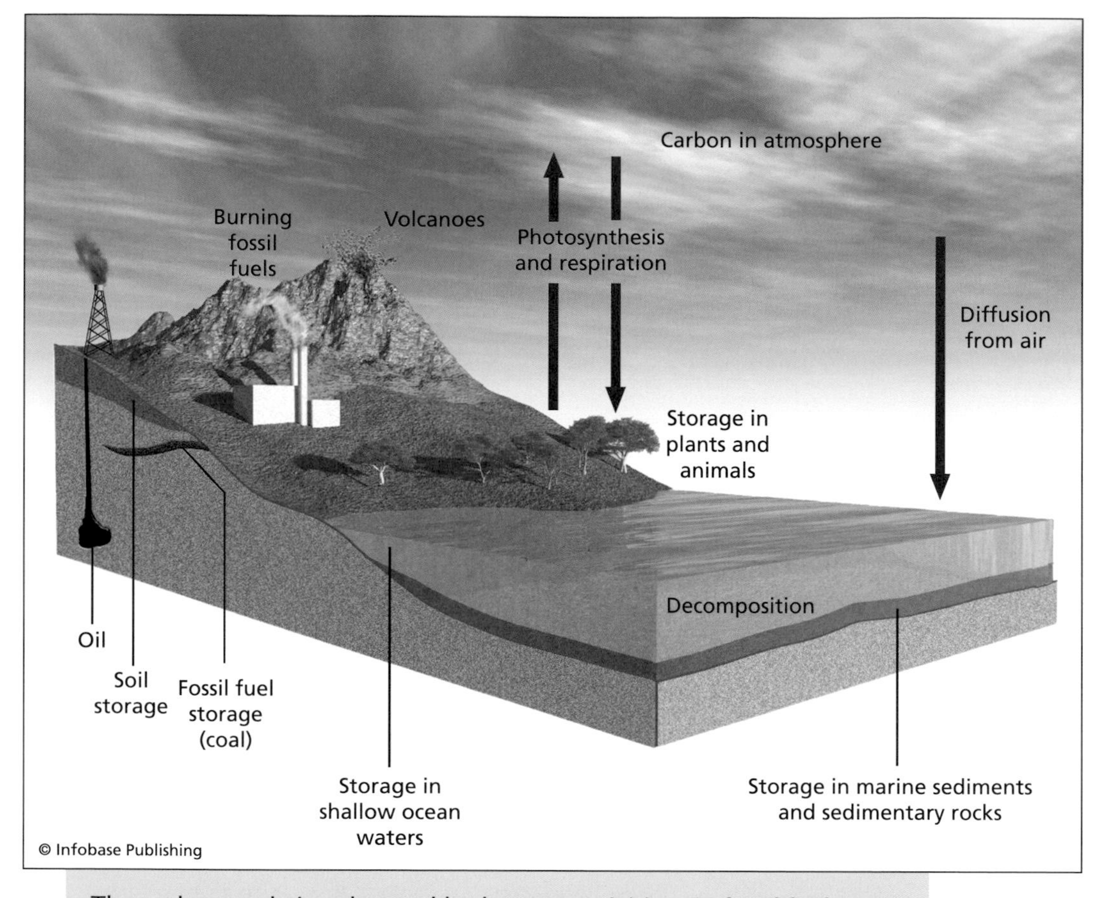

The carbon cycle is enhanced by human activities. As fossil fuels are burned for transportation, industrial processes, heating and cooling of homes, farming, deforestation, and other activities, additional carbon is released into the atmosphere, upsetting the natural balance. *(Source: IPCC)*

In the residential and commercial sector, the main source of direct CO_2 emissions is due to the burning of natural gas and oil for the heating and cooling of buildings. The transportation sector is the second-largest source of CO_2 emissions, primarily due to the use of petroleum products such as gasoline, diesel, and jet fuel. Of this, 65 percent of the CO_2 emissions are from personal automobiles and light-duty trucks.

Specific commodities in the industrial sector that produce CO_2 emissions include cement and other raw materials that contain calcium carbonate that is chemically transformed during the industrial process, which produces CO_2 as a by-product. Industrial processes that use petroleum products also contribute CO_2 emissions to the atmosphere. This includes the production of solvents, plastics, and lubricants that have a tendency to evaporate or dissolve over a period of time. The EPA has identified four principal industrial processes that produce significant CO_2 emissions:

- production of metals such as aluminum, zinc, lead, iron, and steel;
- production of chemicals such as petro chemicals, ammonia, and titanium dioxide;
- consumption of petroleum products in feedstocks and other end uses;
- production and consumption of mineral products such as cement, lime, and soda ash.

Also important are industrial processes that produce other harmful greenhouse gases as by-products, such as methane, nitrous oxide, and fluorinated gases. These gases have a far greater global warming potential (GWP) than CO_2. Methane's global warming potential is 23, nitrous oxide is 296, and fluorinated gases range from 6,500 to 8,700, posing an even more critical negative effect.

TYPES OF CARBON SEQUESTRATION

Carbon sequestration is the process through which CO_2 from the atmosphere is absorbed by various carbon sinks. Principal carbon sinks include agricultural sinks, forests, geologic formations, and oceanic sinks. Carbon sequestration and storage (CSS) occurs when CO_2 is absorbed by trees, plants, and crops through photosynthesis and stored as carbon in biomass, such as tree trunks, branches, foliage, and roots, as well as in the soil. For a more in-depth discussion of the specifics of actual carbon sequestration storage methods, refer to chapter 8.

In terms of global warming and impacts to the environment, sequestration is very important because it has a large influence on levels of CO_2 in the atmosphere. According to the IPCC, carbon sequestration by forestry and agriculture alone significantly helps offset CO_2 emissions that contribute to global warming and climate change.

The amount of carbon that can be sequestered varies geographically and is determined by tree species, soil type, regional climate, type of topography, and even the type of land-management practice used in the area. For example, in agricultural areas, if conservation tillage practices are used instead of conventional tillage, this limits the introduction of CO_2 into the atmosphere by sequestering larger amounts of CO_2 in the soil. According to the EPA, switching from conventional to conservation tillage can sequester 0.11–0.33 tons (0.1–0.3 metric tons) of carbon per acre per year.

Carbon sequestration does reach a limit, however. The amount of carbon that accumulates in forests and soils will eventually reach a saturation point at which no additional carbon will be able to be stored. This typically occurs when trees reach full maturity or when the organic matter contained in soils builds up.

According to the EPA, the U.S. landscape currently functions as an efficient carbon sink, sequestering more than it emits. They do warn, however, that the overall sequestration amounts are currently declining because of increased harvests, land use changes, and maturing forests.

Regarding global sequestration, the IPCC has estimated that 110 billion tons (100 billion metric tons) of carbon over the next 50 years could be sequestered through forest preservation, tree planting, and improved, conservation-oriented agricultural management. In the United States, Bruce McCarl (professor of agricultural economics at Texas A&M University) and Uwe Schneider (assistant professor of the Research Unit Sustainability and Global Change Department of Geosciences and Economics at Hamburg University in Germany) have determined that an additional 55–165 million tons (50–150 million metric tons) of carbon could be sequestered through changes in both soil and forest management, new tree planting, and biofuel substitution.

Another positive aspect supporting carbon sequestration is that it also affects other greenhouse gases. In particular, methane (CH_4) and nitrous oxide (N_2O) can also be sequestered in agricultural activities such as grazing and the growing of crops. Nitrous oxide can be introduced via fertilizers. Instead of using these fertilizers, which can have a negative effect environmentally, other practices could be used instead, such as rotational grazing. In addition, if forage quality is improved, livestock methane emissions should be significantly reduced. Nitrous oxide emissions could be avoided by eliminating the need for fertilizer. The EPA stresses that finding the right sequestration practices will help lessen the negative effects of all the greenhouse gases.

Other environmental benefits of carbon sequestration are that they enhance the quality of soil, water, air, and wildlife habitat. For instance, when trees are planted, they not only sequester carbon, they also provide wildlife habitat. When the rain forests are preserved, they keep both plant and animal species from becoming endangered and help control soil erosion. When forests are maintained they also cut down on overland water flow, soil erosion, loss of nutrients, as well as improving water quality.

The continuation of global warming, however, can have an impact on carbon sequestration. According to a 2001 report issued by the National Academy of Sciences, "Greenhouse gases are accumulating in the Earth's atmosphere as a result of human activities, causing surface air temperatures and subsurface ocean temperatures to rise." Besides temperature, they also say that human-induced climate change may affect the growing seasons, precipitation patterns and amounts, as well as the frequency and severity of extreme *weather* events such as the wildfires that currently plague the American Southwest.

According to the EPA: "In terms of global warming impact, one unit of CO_2 released from a car's tailpipe has the same effect as one unit of CO_2 released from a burning forest. Likewise, CO_2 removed from the atmosphere through tree planting can have the same benefit as avoiding an equivalent amount of CO_2 released from a power plant."

The experts at the EPA also caution, however, that even though forests, agriculture, and other sinks can store carbon, the process can also become saturated and slow down or stop the storage process (such as traditional agricultural cultivation), or the sink can be destroyed and completely stop the process (such as complete deforestation). Carbon sequestration processes can naturally slow down and stop on their own when they get older.

In addition, carbon sequestration can be a natural or man-made process. Research is currently in progress to perfect the methodologies that enhance the natural terrestrial cycles of carbon storage that remove CO_2 from the atmosphere by vegetation and store CO_2 in biomass and soils. In order to accomplish this, research of biological and ecological processes are under way by the EPA. Specific technical areas that are currently being researched include:

- Increasing the net fixation of atmospheric CO_2 by terrestrial vegetation with emphasis on physiology and rates of photosynthesis of vascular plants;
- Retaining carbon and enhancing the transformation of carbon to soil organic matter;
- Reducing the emission of CO_2 from soils caused by heterotrophic oxidation of soil organic carbon;
- Increasing the capacity of deserts and degraded lands to sequester carbon. Man-made processes include technologies such as geologic, mineral, and ocean sequestration.

In carbon sequestration, the main goal is to prevent CO_2 emissions from power plants and industrial facilities from entering the atmosphere by separating and capturing the emissions and then securing and storing the CO_2 on a long-term basis.

Currently, the EPA is involved in research in an attempt to separate and capture the CO_2 from fossil fuels and from flue gases—both pre- and post-combustion processes. Underground storage facilities are also receiving large amounts of attention recently, and their potential is enormous. As an example, today more than 750 billion gallons (2.8 trillion liters) of both hazardous and nonhazardous fluids are disposed through a process called "underground injection."

There are certain risks associated with underground sequestration, however. Leakage from the storage reservoir is a concern that must be taken seriously when selecting the right location for storage. Because this is a relatively new technology, much still needs to be learned about appropriate safety measures. The potential risks of underground CO_2 sequestration include escape of CO_2 from the reservoir through leakage, seismicity (shifting of the Earth's crust from seismic activity), ground movement, or displacement of brine (if present). Leakage is the largest of the risks, which could occur through or along abandoned wells and by cap rock failure. A cap rock is the "impermeable" rock formation above the deposit, keeping it naturally contained.

Another issue to consider is that diffusion of CO_2 through the cement or steel casing it is contained in—caused by corrosion—may be a slow process. However, when it is sequestered for tens of thousands of years, the integrity of the well casing needs to be critically assessed. Other areas of potential concern are through high permeability zones in the cap rock above the deposit or through faults and fractures that extend into the cap rock. Those areas are considered the most important natural leakage pathway.

Agricultural Sequestration

Carbon sequestration occurs in soils and agricultural crops mainly through the natural process of photosynthesis. Half of the agriculture biomass is composed of carbon. When the vegetation decays, the litter and roots also contribute carbon to the soil. When agricultural fields are plowed, CO_2 can be released to the atmosphere. The amount of carbon in cropland soils can vary depending on types of crops, types of fertilizers used, and type of management practice (such as conventional or conservation tillage). These variables can also determine the presence of other greenhouse gases such as methane and nitrous oxide, which have greater global warming potentials than CO_2.

According to the EPA, carbon can be sequestered in agricultural soils for 15 years or longer depending on the soil type and management style. There are several agricultural practices that sequester carbon and also reduce the emissions of other greenhouse gases (such as nitrous oxide and methane). According to the EPA, the following table depicts the methods in which agriculture can sequester carbon.

With changes in grazing methods, land management can sequester carbon for 25 to 50 years. According to Tristram O. West of the Department of Energy (DOE) Center for Carbon Sequestration in Ecosystems (CSITE) at Oak Ridge Laboratory in Tennessee, using reduced tillage

Agricultural Practices That Sequester Carbon or Reduce the Emissions of Other Greenhouse Gases

KEY AGRICULTURAL PRACTICES	TYPICAL DEFINITION AND SOME EXAMPLES	EFFECT ON GREENHOUSE GASES
Conservation or riparian buffers	Grasses or trees planted along streams and croplands to prevent soil erosion and nutrient runoff into waterways.	Increases carbon storage through sequestration.
Conservation tillage on croplands	Typically defined as any tillage and planting system in which 30 percent or more of the crop residue remains on the soil after planting. This disturbs the soil less, and therefore allows soil carbon to accumulate. There are different kinds of conservation tillage systems, including no till, ridge till, minimum till, and mulch till.	Increases carbon storage through enhanced soil sequestration, may reduce energy-related CO_2 emissions from farm equipment, and could affect N_2O positively or negatively.
Grazing land management	Modification to grazing practices that produce beef and dairy products that lead to net greenhouse gas reductions (such as rotational grazing).	Increases carbon storage through enhanced soil sequestration and may affect emissions of CH_4 and N_2O.
Biofuel substitution	Displacement of fossil fuels with biomass (agricultural and forestry wastes, or crops and trees grown for biomass purposes) in energy production, or in the production of energy-intensive products such as steel.	Substitutes carbon for fossil fuel and energy-intensive products. Burning and growing of biomass can also affect soil N_2O emissions.

Source: EPA

Carbon is sequestered in the crops and soil of agricultural areas. *(U.S. Department of Agriculture)*

methods instead of conventional methods can sequester carbon for 15 to 20 years.

A major benefit of agricultural sequestration is that it also positively affects soil, air, water, and therefore, wildlife habitat. When carbon sequestration practices reduce soil erosion and management practices decrease pollution runoff, this all promotes healthy water quality in addition to preventing climate change.

Forestry

Forests accumulate carbon both above and below ground as well as in soil pools. The amount of carbon sequestered is determined by both soil and vegetation type. Carbon can be sequestered for decades, even centuries, until the forest reaches its point of saturation, unless the carbon is released prematurely through forest fire, natural decay, or harvesting.

According to the EPA, the following table outlines forestry practices that sequester carbon.

Based on the data collected by the EPA, U.S. forests and croplands currently sequester more than 600 teragrams (Tg) of CO_2 equivalent. This equals 170 Tg or million metric tons of carbon equivalent. This is an amount equivalent to 12 percent of the total U.S. CO_2 emissions from the energy, transportation, and industrial activities of the nation.

Forestry Practices That Sequester Carbon

FORESTRY PRACTICES	DEFINITION AND EXAMPLES	EFFECT ON GREENHOUSE GASES
Afforestation	Tree planting on lands previously not part of forests, such as the conversion of cropland to trees.	Increases carbon storage through sequestration.
Reforestation	Planting trees on lands that used to be forested, such as replanting areas burned by forest fires.	Increases carbon storage through sequestration.
Forest preservation or avoided deforestation	Protection of forests that are threatened by logging or clearing.	Avoids CO_2 emissions through the conservation of existing carbon stocks.
Forest management	Modifying forest practices that produce wood products to prolong sequestration of carbon, such as not harvesting as often.	Increases carbon storage by sequestration. Also avoids CO_2 emissions by changing land-management practices.

Source: EPA

One of the biggest contributors to upsetting the natural carbon balance is deforestation, especially of the world's rain forests. This photo shows deforestation presently occurring in Madagascar. Once deforested, hillsides are vulnerable to erosion. *(Rhett Butler, Mongabay.com)*

On a worldwide level, roughly 20 percent of the world's annual CO_2 emissions originate mainly from deforestation in the world's tropical regions of South and Central America, Asia, and Africa. This represents a huge input of CO_2 into the atmosphere in return for alternative uses of the land. Generally, the high-carbon biomass is removed in the trees, and the land is then used for crops or grazing (these activities store much less carbon than forests).

A long-term storehouse of carbon is in the world's forests. Carbon will remain stored in the biomass (trees, shrubs, and other vegetation) until it dies, is burned, or is purposefully cut down. *(Nature's Images)*

If the United States wants to further offset/reduce CO_2 emissions—such as by the 7 percent stated in the Kyoto Protocol—by planting trees, it would require planting an area as big as the state of Texas every 30 years. The Kyoto Protocol is an international agreement linked to the United Nations Framework Convention on Climate Change (UNFCCC). The protocol sets binding targets for 37 industrialized countries and the European community for reducing greenhouse gas emissions. It was adopted in Kyoto, Japan, on December 11, 1997, and came into force on February 16, 2005. To date, 184 countries have ratified the protocol. The United States, although a signatory to the protocol, has neither ratified nor withdrawn from the protocol, making it nonbinding. The reason

President George W. Bush did not submit the treaty for Senate ratification was because both China and India were considered undeveloped countries and held exempt from greenhouse emission control. China is currently the world's largest gross emitter of CO_2.

Geologic Sequestration

CO_2 can also be captured at stationary areas and injected underground into geological formations—a process called geological sequestration. This technology captures the CO_2 and other pollutants emitted when fossil fuels are burned. They are then compressed into a liquid and pumped deep beneath the Earth's surface into geologic formations. David Goldberg, a geophysicist at Columbia University's Lamont-Doherty Earth Observatory in Palisades, New York, has experimented successfully with this technique by pumping CO_2 into basalt rock beneath the ocean floor. This is one method the IPCC has identified to mitigate global warming. The IPCC has suggested that offsetting CO_2 emissions in this manner is one way to achieve an overall net reduction in greenhouse gas emissions. The IPCC has also estimated that worldwide it is possible to permanently store up to 1,100 gigatons of CO_2 underground. As a comparison, worldwide emissions from large stationary sources such as urban areas is about 13 gigatons per year. Geological sequestration could store approximately 85 years of CO_2 emissions from large stationary stores.

The IPCC endorsed this sequestration technology as safe and acceptable in 2005 because CO_2 has been sequestered naturally in geologic formations for hundreds of millions of years with no ill effects. The IPCC has also supported the technology because there are several practical trapping mechanisms available that can keep the CO_2 from moving around underground and eventually escaping.

According to the EPA, about 95 percent of the largest sources of CO_2 emissions (coal-fired power plants, large cities) in the United States are within 50 miles (80 km) of a potential geologic sequestration site. In order to pursue this technology, the EPA is currently investigating geologic sequestration technology.

The method—also called geo-sequestration or geological storage—involves injecting CO_2 directly into underground geological formations.

Storage sites that would qualify include old oil fields, unmineable coal seams, and saline aquifers.

The option of using old oil fields is an attractive option because CO_2 has been injected into existing fields for the past 30 years in order to recover the last of the recoverable oil. Besides having already been done extensively and with the knowledge that the technology works, an additional benefit with these sites is that these formations already have an impermeable cap rock above them to keep the CO_2 from escaping into the atmosphere (the same barrier that kept the oil underground for millions of years instead of leaking out onto the Earth's surface).

Unmineable coal seams are also considered to be safe and able to supply long-term storage. Saline aquifers are another formation considered as a possibility to store CO_2 for a long period of time. According to the EPA, they have a large potential storage area associated with them, and they are relatively common geographically, which means that the CO_2 would not have to be transported long distances, thereby lowering sequestration costs. The downside to saline storage sites is that they have not been used enough at this point to fully understand their positive or negative characteristics yet. Current research is being conducted in southeastern Saskatchewan, trying to determine whether the pros outweigh any cons.

According to a report in iStockAnalyst and *The Paducah Sun*, three energy companies along with the Kentucky Geological Survey are planning to test the feasibility of storing CO_2 underground. The Western Kentucky Carbon Storage Foundation will be providing the resources to drill the test well. Peabody Energy (the world's largest coal company) and ConocoPhillips (the world's fifth-largest oil refiner) are looking into the possibility of building a coal gasification plant to produce a substitute for natural gas—the CO_2 sequestration project is part of this larger project.

Rick Bowersox, chief geologist for the Kentucky Geological Survey, says, "We want to demonstrate the sequestration process prior to it being put into commercial use."

In response, Kentucky governor Steve Beshear says, "The importance of CO_2 sequestration research for our state's and the nation's energy future cannot be overstated. Kentucky has been a strong supporter of effects to develop partnerships and projects such as this one, and the results from this testing are crucial in our quest to develop and implement clean coal technologies and become a world leader in energy production."

Australia has also announced recent plans to try carbon sequestration in order to offset carbon emissions. Prime Minister Kevin Rudd announced in July 2008 a $5 billion plan to implement carbon sequestration by storing CO_2 in deep subterranean strata and the adjacent seabed. Although he has been met with criticism, saying that carbon sequestration technology is too new and untried to invest that much money in trying, Rudd is an optimistic supporter of the technology, and he plans to move ahead.

Resources Minister Martin Ferguson agrees that there are "good arguments for implementing carbon sequestration." He believes it would encourage investment and commercialization of the technology and is a feasible way to allow the continued generation of carbon-based power while reducing the net overall impact.

Currently, the Commonwealth body Geoscience Australia has identified several areas that would be feasible for storage of greenhouse gases—areas in Victoria, Western Australia, and Southern and Central Queensland. As the project moves forward, however, it is meeting opposition from bureaucrats claiming that up to 15,000 jobs in the energy and production industries that do generate greenhouse gases could be lost.

Also reported in July 2008 by *Scientific American News,* volcanic rocks buried deep beneath the ocean off the coast of the Pacific Northwest (California, Oregon, and Washington) may prove to be one of the best places yet to store CO_2.

David Goldberg, a geophysicist at Columbia University's Lamont-Doherty Earth Observatory, says that pumping greenhouse gases deep into basalt rock in the region is a workable solution to offset increasing amounts of CO_2 in the atmosphere. The volcanic rock has proven desirable because at 8,850 feet (2,700 m) below the ocean surface the liquid CO_2 is heavier than the water above it, ensuring that leaks would never be able to bubble to the surface. In addition, CO_2 mixes with the volcanically warm water and undergoes a chemical reaction within the basalt itself to form carbonate compounds called chalk, which permanently locks up the greenhouse gases into a mineral form. For deposits in the ocean bed, the additional layer of 650 feet (200 m) of marine sediment on top of the basalt layer acts as another barrier.

According to Goldberg, "You have three superimposed trapping mechanisms to keep your CO_2 below the sea bottom and out of the atmosphere. It's insurance on insurance on insurance."

In this project alone, it is estimated that at least 229 billion tons (208 billion metric tons) of carbon could be stored in the basalt formations of the Juan de Fuca Plate (a major plate in the Earth's plate tectonic system) 100 miles (160 km) off the United States's West Coast. This would represent the equivalent of 122 years of all U.S. emissions of 1.9 billion tons (1.7 billion metric tons) each year.

M. Granger Morgan of Carnegie Mellon University, who is a carbon sequestration expert, says, "On the one hand you wouldn't want to bet the future of U.S. climate policy on it until one has done more work, but on the other hand, it looks quite promising."

Goldberg is also looking at the issue globally and says there are other formations worldwide similar to this tectonic plate in the Pacific Northwest that could be used for CO_2 sequestration. One promising site in Iceland began testing in 2008. The project involves the capture and separation of fuel gases at the Hellisheidi Geothermal Power Plant, the transportation and injection of the CO_2 gas fully dissolved in water at elevated pressures at a depth between 1,312–2,625 feet (400–800 m), as well as the monitoring and verification of the storage. A comprehensive reservoir characterization study, currently in progress, must be completed before the CO_2 injection can occur. This includes soil CO_2 flux measurements, geophysical survey, and tracer injection tests. The results from the tracer tests are showing significant tracer dispersion within the target formation, which suggests that there is a large surface area available for chemical reactions. The pilot project's large available reservoir volume and surface area in combination with the relatively fast CO_2-water-rock reactions that occur in basaltic rocks look like they may allow safe and permanent geologic storage of CO_2 on a large scale. According to Goldberg, "The volumes of storage we're talking about are huge and the problem is huge. The prize is very large here with this option. It's worth a serious look."

The Midwest Regional Carbon Sequestration Partnership claims that a porous rock filled with saltwater that underlies much of the Midwest could permanently store half of the greenhouse gases released in the next century by industries in Ohio and neighboring states. Because of this, they are currently involved in a large-scale test injection of CO_2 into the rock layer, known locally as the Mount Simon Sandstone formation.

The goal of this project is to compress and inject 1 million tons (907,185 metric tons) of CO_2 from a new ethanol plant in Greenville. The gas would be injected into a chamber to a depth of 3,000 feet (914 m) from 2010 to 2014.

This project would offset the Andersons Marathon Ethanol LLC—Ohio's largest ethanol plant—which will produce 110 million gallons of ethanol from 43 million bushels of corn each year. Generating more than 250,000 tons of CO_2, almost the entire amount of CO_2 would be injected underground.

Debra Crow, spokesperson for Andersons Marathon Ethanol LLC, says, "We're interested in being a good corporate citizen and helping with any kind of research that can improve the environment."

The U.S. Department of Energy is providing $61 million in funding for the Greenville project. Not everyone sees it in a positive light, however. According to Nechy Kanter of the Sierra Club, "The government should place a higher priority on improving energy efficiency and supporting green technologies such as wind and solar. Carbon sequestration is 'highly experimental and highly expensive,' and can't be undertaken without taxpayer support. We don't want polluting sources counting on sequestration and then it not come through."

Mineral Sequestration

The process of mineral sequestration traps carbon in the form of solid carbonate salts. This type of sequestration is very slow. An example of this in nature is the deposition and accumulation of limestone (calcium carbonate) over geologic time. It is the carbonic acid found in groundwater that slowly reacts with the silicates to dissolve magnesium, calcium, silica, and alkalis to form clay minerals. When creatures with shells die, the shells are deposited in the sediments on the seafloor and

THE CARBON CONUNDRUM

According to experts at NASA's Goddard Space Flight Center, a critical aspect of global warming is the result of temperature change on carbon storage in the biosphere. NASA scientists have discovered an interesting phenomenon. They have discovered that an increase in global temperatures may both increase and decrease the rise in atmospheric CO_2.

This is their discovery: When temperatures rise, carbon is released from the soil; when it oxidizes (joins with oxygen), it is converted into CO_2. When temperatures rise, there is also a notable increase in plant growth, which in turn absorbs CO_2, taking it out of the atmosphere. In this way, an ecosystem has the ability to help moderate the balance of CO_2 concentrations present between global warming and plant growth.

David Schimel of the Climate and Global Dynamics Division of the National Center for Atmospheric Research in Colorado says, "Within the literature, there's been a debate about the effect of temperature on ecosystems. One line of argument holds that as temperatures increase, ecosystems should lose carbon to the atmosphere (because the temperature/carbon release effect is stronger than the increased temperature/plant growth effect). The opposing argument holds that most warm weather carbon release is from the soils. One catch, however, is that when carbon is released from the soils, so are nutrients, which then act as a fertilizer on the vegetation, which should increase plant growth and increase photosynthesis, thereby taking in more CO_2 from the atmosphere.

Whereas both theories should hold true, Schimel conducted a study. He concluded that "the data support both hypotheses, but the difference between the two relates to an offset in time: (1) in warm years there is a release of carbon. Then, one to three years later, there is some carbon uptake." According to Schimel, the next problem for NASA to address is which effect is larger.

He has determined that at the peak of warm periods, plant growth increases in both the polar and temperate regions, but decreases in the equatorial regions. Conversely, in global warming conditions, research suggests decreased plant growth. Schimel believes the future role of models that analyze the atmosphere and biosphere together will be crucial.

eventually become transformed into layers of limestone. Limestone can accumulate over billions of years of geologic time. Because these formations are so high in carbon, large stores of the Earth's carbon are located in these formations.

Based on this slow, natural process, researchers are trying to find artificial ways to speed up this process. Currently, one method under consideration is to take dunite or serpentinite and mix it with CO_2 to form the carbonate mineral magnesite, plus silica and iron oxide (also called magnetite).

So far, serpentinite sequestration is the preferred method because it is nontoxic and a more stable, predictable rock. The drawback in this experimental method is that the serpentine works best only if the formation has a high magnesium content. If iron is present in the formation, this method is not as attractive.

Ocean Sequestration

Another strategy proposes to store CO_2 in the world's oceans. Several methods have been suggested. One option is "direct injection." In this method, CO_2 is pumped directly into the water at extreme depths of 1,148–11,811 feet (350–3,600 m). The CO_2 forms a mass on the seabed and transforms into solid CO_2 clathrate hydrates, which gradually dissolve into the water around it. There are problems associated with this method, however. It could contribute to ocean acidity and could lead to increased levels of submerged methane gas, which if it reached the surface and escaped into the atmosphere, would also cause a serious greenhouse gas problem.

Another suggested method of long-term sequestration is to bale large bundles of agricultural crop waste biomass (such as corn stalks, hay, and grass) and deposit it into alluvial fans in deep ocean basins, where it would eventually get buried under thick layers of sediment. Critics of this method caution that there could be an increase in aerobic bacteria growth, which would lower the available oxygen levels to ocean life-forms. It has also been suggested to convert CO_2 to bicarbonates, using limestone or to store CO_2 in solid clathrate hydrates already residing on the ocean floor.

Ocean sequestration is still in its infancy, and much more needs to be learned about it until it can become a proven environmentally sound method of sequestration. To date, pumping CO_2 into deep seafloor geological formations appears to be the best type of "ocean sequestration" such as that proposed in the Pacific Northwest, discussed earlier.

According to a report in the *Seattle Times,* the CO_2 would be pumped into porous basalt layers up to a half-mile below the ocean bottom. Dave Goldberg, when talking about the CSS storage in the Pacific Northwest mentioned previously, stated, "This is the first good example of a site that is of the scale that can potentially make a dent on the problem of CO_2 storage."

The minerals in the bedrock would serve to form stable carbonates. The deposits would also be covered by sediments 1,000 feet (305 m) thick to eliminate the chance of leaks. It is also a convenient location: close enough to the shoreline that the CO_2 could be piped directly from power plants to injection sites.

According to Taro Takahashi, senior scholar of Lamont-Doherty, "Undersea storage has been largely ignored, even though there's much more capacity, and repositories would be much less prone to leak. In principle, the type of reservoir we propose at the ocean floor is one of the safest—if not the safest—way of storing liquefied CO_2 for a long, long time."

Some argue, however, that there may be unknown ecological impacts and seismic hazards associated with the technology.

Other Sequestration Methods

Currently, there is a small company in the United States that is working on a new technology designed to sequester CO_2. The method involves taking CO_2 along with tailings from mining operations, combining them, and converting them into a new material that can be used in a variety of industrial, agricultural, and environmental applications. The new material has been dubbed "green carbon."

Based on a report in *Environmental News Network,* a company called Carbon Sciences has developed a relatively simple technology that puts the mixture under pressure and temperature to create "precipitated calcium carbonate" (PCC). PCC is a common component of many products used everyday such as paper, plastic, wallboard, food additives, pharmaceuticals, vinyl siding, fencing, agricultural products, and fertilizer.

The traditional way of processing calcium carbonate is to put it through an energy-intensive process that is cost-prohibitive because it uses expensive materials such as limestone. This new "green carbon" technology, however, simplifies the traditional process and is able to not only sequester CO_2, but also to produce a usable material from it.

Carbon Sciences, founded by Derek McLeish, points out that when a "green carbon technology" is applied to a "carbon neutral process," such as an ethanol plant, that process actually reduces the amount of CO_2, making the overall operation "carbon negative." Therefore, McLeish states that "not only is Green Carbon a method for removal and transformation of CO_2, but the technology can produce PCC at a lower cost than traditional processes." In addition, he believes that as carbon credit markets come online, users will automatically realize additional cost reductions when the CO_2 consumed in green carbon is sold as carbon credits. Carbon credit is a concept that allows carbon conservation in one area to be applied as a counterbalance toward the effects of carbon pollution in another area. For example, carbon credits may be applied if a fossil fuel burning company plants trees elsewhere (which take in and store CO_2 long-term) in order to even out the negative effects of adding CO_2 to the atmosphere through the burning of fossil fuels.

Potential Sequestration Drawbacks

One study that appeared in Science*Daily* in July 2008 examined different methods of carbon sequestration in an attempt to determine the pros and cons to the environment using four different testing scenarios, which conserved different types of land and animal species. The study determined that conservation of species is usually maximized when the natural habitat is either restored or maintained. Carbon sequestration, however, is maximized when forest ecosystems are restored.

According to Andrew Plantinga, a professor in the Department of Agricultural and Resource Economics at Oregon State University, "The main thing we found is that if you want to conserve species, that policy might not be compatible with carbon sequestration. On the other hand, if you want to get carbon out of the atmosphere, it's not clear that will be good for the species. The take-home message is this: when you think about policies targeted to private landowners, government has to be careful about how it does this because it may achieve one objective but at the expense of something else."

Another drawback is the cost associated with carbon sequestration. Costs depend on the method used. Sometimes, when declining oil fields are used to store carbon, the oil withdrawn from the field during the sequestration can offset the costs. Recovering the oil in this option, however, reduces the environmental benefit of storing the carbon because the carbon in the oil being withdrawn offsets the environmental benefits from storing the carbon.

MODELING CARBON EMISSIONS IN THE UNITED STATES

A study conducted by Purdue University has recently shown that the CO_2 emissions in the United States are not all located where they were expected to be. According to Kevin Gurney, assistant professor of earth and atmospheric science at Purdue University, "We've been attributing too many emissions to the northeastern United States, and it's looking like the southeastern United States is a much larger source than we had estimated previously."

Purdue's new CO_2 modeling system, called Vulcan, develops models more than a hundred times more detailed than prior models. They are collecting CO_2 emissions at local levels on an hourly basis. Prior to that, readings were collected only once a month on a statewide basis. One of the strong points about this new model is the derivation of the CO_2 value. The CO_2 reading is calculated by transforming the data on local air pollution (such as carbon monoxide and nitrous oxide).

It is the hope of the project team that Vulcan will be able to help lawmakers create effective policies to reduce CO_2 emissions in appropriate amounts in the right areas.

According to Gurney, "Before now, the only thing policy makers could do was take a big blunt tool and bang the U.S. economy with it. Now we have more quantifiable information about what is happening in neighborhoods, on roads, and in industrial areas, and track the CO_2 by the hour. This offers policy makers something akin to a scalpel instead."

Vulcan's CO_2 values reflect all the CO_2 from the burning of fossil fuels and tracks hourly outputs from power plants, industries, highways, business districts, and residential neighborhoods. As soon as Vulcan produced its maps, the research team realized that past CO_2 calculations were inadequate. Gurney further adds, "When you compare the old inventories to Vulcan, the new data show atmospheric CO_2 differences that are as large as five parts per million in some U.S. regions in the late winter. The levels in the global atmosphere only rise one and a half parts per million every year, so this is the equivalent of three years of global emissions in the atmosphere that isn't where we thought it was. This will be important for policy makers and is enormous from a scientific point of view. It's shocking."

According to James E. Hansen at NASA's Goddard Institute for Space Studies, Vulcan is helpful because it provides a "check to judge the accuracy of existing satellite data."

Robert Andres of the Carbon Dioxide Information Analysis Center at the U.S. Department of Energy's Oak Ridge National Laboratory in Tennessee says the Vulcan project is able to provide scientists with many insights into the global carbon cycle and climate change. He also commented, "Vulcan will be revolutionary in carbon cycle research. It is the next generation in our understanding of fossil fuel emissions. The implications for climate science, carbon trading, and climate change mitigation work are tremendous."

Models like Vulcan are now being developed that allow scientists to understand the complex issues at hand with global warming and climate change. If scientists can convey the message in a meaningful way to policy makers such as the Vulcan updateable interactive mapping system, the education process of global warming can be put into motion. Then, if policy makers understand the critical importance of acting now and begin measures to cut carbon output and sequester existing carbon sources, slowing the ill effects of global warming can become a reality that will benefit future generations.

Agriculture and Greenhouse Gases

Global warming will have an effect on agriculture worldwide, but the outcomes will not be consistent. There may be positive effects in some areas and negative ones in others. As global warming influences temperatures and precipitation, it will play an important role in what types and amounts of crops can be produced. This will have a crucial effect on the world's food supply—what can be grown, how much, and where. This chapter explores both the positive and negative effects of global warming on agriculture. It then illustrates how agriculture functions as a natural resource and how it can assist the environment in combating global warming by being used as a source of mitigation against greenhouse gases. Next, it focuses on issues of human *adaptation,* food supply, and what future impacts are expected as a result of continued global warming.

THE EFFECTS OF GLOBAL WARMING ON AGRICULTURE

One of the most important resources society depends upon is agriculture. With so much irrefutable evidence demonstrating that global warming is indeed occurring and that ecosystems worldwide are already feeling the effects and having to adapt, the protection of agricultural resources is critical. Without the successful adaptation of agriculture and livestock to the long-term changes of climate due to global warming, future adaptation for anything would be difficult. Agricultural systems are influenced by several environmental factors—especially weather and climate. Agriculture and livestock depend on the health and well-being of soil conditions such as the presence and quality of organic matter and availability of adequate moisture. If the progression of global warming upsets the balance of any of these biophysical properties (precipitation, temperature, soil moisture, and organic matter), agriculture and livestock will be negatively affected.

Currently, much of the work done to understand global warming and determining how specific areas may change is through computer modeling. Modeling climate change is very complicated, however. Precipitation patterns and cloud formation and movement are currently very difficult to model because clouds are small-scale features that move and change rapidly; modeling is most successful on the mechanisms of climate that occupy larger scales and move slower, such as massive air masses like the global circulation patterns of jet streams, westerlies, and *trade winds.*

Biophysical factors such as soil conditions and type and amount of organic matter help determine soil moisture and crop conditions. Climate change also reflects these ranges of conditions, which in turn affects crop production. Because of this, extensive research is conducted at the U.S. Department of Agriculture on developing new varieties of crops that are geared toward reducing their vulnerability to climatic stress, as well as expanding the range of geographic, topographic, and climatic conditions in which they can be successfully grown and harvested.

THE RESPONSE OF AGRICULTURE TO CLIMATE CHANGE

According to the Pew Center on Global Climate Change, agricultural production will experience a mixture of positive or negative effects that will depend on the geography and specific conditions of the area. The

Pew Center, established in 1998, is a nonpartisan, independent research organization. The center's mission is to provide credible information, straight answers, and innovative solutions in the effort to address global climate change. Future predictions from global warming models have predicted higher temperatures as well as higher precipitation in certain areas.

The effect of increased temperature can have either a positive or negative effect, based on geography. For instance, for the colder high-latitude locations such as Alaska and Canada, warmer temperatures would enable the existence of longer growing seasons. This could encourage the possibility of extended growing seasons as long as soil conditions are adequate. If soil conditions are not adequate and are not fertile enough (lacking proper nutrients and soil structure) to support the growth of crops, the length of the growing season will not matter because growing crops would not be successful. Adding large doses of fertilizer to increase fertility is also not a good option because they can have negative effects on the environment, such as being washed off into the drainage and entering and polluting the water cycle and biogeochemical systems.

In more southern latitude locations, warmer temperatures would be beneficial if they eliminated the killing frosts that can currently destroy crops. This could potentially enable citrus crops to be grown farther north. The U.S. Department of Agriculture warns, however, that high temperatures could shorten the critical growing period interval, limiting the successful growth of large crop yields.

Changes in precipitation may also have positive and negative effects. Increased precipitation in the traditionally arid areas such as the American Southwest would benefit them by enabling growing periods to exist. On the other hand, precipitation increase in areas where it results in flooding would be a disaster, causing erosion and loss of land and resources. In other geographical locations that experience a shortfall of precipitation, drought and water shortage would cause havoc with agricultural production.

Increasing levels of CO_2 in the atmosphere usually increases the net photosynthetic rates in plants, as well as reduces transpiration loss. Because the photosynthesis rate is enhanced in plants as atmospheric

CO_2 increases, vegetation productivity rises. The specific reaction to increased CO_2 to each plant varies by plant species. Based on findings at the Pew Center, the main commercial crops grown in the United States are wheat, rice, potatoes, barley, oats, and a wide variety of vegetables, which typically respond well to an increase in CO_2. With a modeled increase of a doubled level of CO_2, crops in the United States would be expected to generally increase quantities by 15 to 20 percent. On the other hand, the tropical vegetation species—or warm weather crops—such as sugarcane, sorghum, corn, and tropical grasses are much less responsive and estimates of only 5 percent are expected.

The difficult aspect of this modeling is that some scientists believe crop productivity is also significantly difficult to predict because CO_2 is also influenced by other things such as nutrients and water availability, which makes it more difficult to predict the future. In computer modeling done by the Intergovernmental Panel on Climate Change (IPCC), the estimated effects of global warming on agricultural resources will vary from crop to crop and geographic region to geographic region, as shown in the following table.

General predictions have been made that areas that experience large temperature changes will experience more negative effects. Countries in the mid- and high-latitude areas may experience an enhanced CO_2 effect accompanied by an increase in agricultural production; but yields in low-latitude countries such as Brazil may actually see a decline in agricultural production, especially with commodities such as corn and wheat, as drought becomes more likely. The commodity that seems to be the most resilient to climate change is rice. According to the IPCC, specific temperature ranges may also play an important role. For instance, areas that increase up to 3.6°F (2°C) are the areas that may experience positive crop yields. Areas that have an increase of 7.2°F (4°C), on the other hand, may actually see less production.

Livestock are also greatly affected by global warming. They can be affected in two major ways: (1) by the quantity of grazable land and whether there is enough to support the number of livestock on it; (2) the quality of grazable land—the types and abundance of forage available for the health of the livestock living on the land. Global warming can seriously affect both of these variables.

Estimated Climate Change Effects on Crop Yields in Latin and North America	
LOCATION OF STUDY SITE	IMPACT (CROP: PERCENT CHANGE IN YIELD)
North America:	
Canada	Wheat: –40% to +234% (results vary due to geographic location)
United States	Corn: –30% to –15% Wheat: –20% to –2% Soybean: –40% to +15%
Latin America:	
Argentina	Corn: –36% to –17% Wheat: +3% to +48% Soybean: –3% to –8%
Brazil	Corn: –25% to –2% Wheat: –50% to –15% Soybean: –61% to –6%
Mexico	Corn: –61% to –6%
Note: Increases are due to cooler areas becoming warmer and more conducive to the growth of crops; decreases are due to areas becoming more arid and less conducive to the growth of crops. *Source: IPCC*	

Another variable that can affect livestock is the encroachment of invasive species and pests. As temperatures warm, invasive plants, grasses, and weeds can overtake rangeland biomes. If these invasive species are not foods that grazing livestock can eat, the food supply is threatened, hence, their survival. Extreme weather can also play a significant role in the health and well-being of livestock. As global warming progresses, it is expected that severe *weather* episodes will increase, such as blizzards, flooding, heat waves, and drought. Each of these

Cattle grazing on open rangeland in the western United States
(Bureau of Land Management)

weather events and climate conditions can have a negative impact as well, as evidenced by what occurred during the 1996–97 winter season in the northern plains of the United States. During that winter, severe blizzards and accompanying freezing temperatures took a devastating toll on the livestock. In the American Southwest, several years of drought from 2001 through 2005 drastically lowered the carrying capacity of the land for the grazing of cattle and sheep. Winter snowpacks in the mountains were unseasonably low, resulting in a low, early spring runoff. Drought conditions caused what little vegetation there was to dry up and die quickly, leaving livestock without forage to graze on. What scarce, brittle vegetation was left was then highly susceptible to wildfire caused by both lightning strikes and human carelessness, making summers prone to high incidences of damage by fire. The summer and fall of 2007 was one of the most devastating periods of wildfire due to dried-out, drought-stricken vegetation.

According to the Pew Center, another undesirable effect related to high temperatures is that under these conditions, livestock's appetites are suppressed, causing lower weight gain. Based on studies conducted by R. M. Adams (Department of Agriculture and Resource Economics, Oregon State University), J. D. Glyer (Christensen and Associates, Madison, Wisconsin), and B. A. McCarl (Texas A&M University), with a 9°F (5°C) increase in temperature cow/calf herd population in the United States dropped by 10 percent.

Another team of scientists—J. D. Hanson (USDA), B. B. Baker (USDA), and R. M. Bourdon (Department of Animal Sciences, Colorado State University)—found that a warming of 2.7°F (1.5°C) causes a 1 percent loss of livestock herds. They also found that the herds were less productive and did not produce as much milk. In this study, the scientists concluded that the negative effects were caused by both the increased temperature and lack of forage.

The Pew Center on Global Climate Change has also identified some indirect effects of climate change on agriculture. One of the most significant is soil erosion and degradation. The effects of flooding, drought, and wind erosion can have devastating effects on crop yields.

Positive Effects of Climate Change

According to a report in *National Geographic News* in November 2002, as greenhouse gas levels climb higher in the atmosphere, crop yields may increase in certain areas, such as fruits, vegetables, seeds, and grains. Although this may initially sound like a good thing, the positive effect is short-lived because it also results in a decrease in the crop's nutritional value.

Peter S. Curtis, an ecologist at Ohio State University, says, "But there's a trade-off between quantity and quality. While crops may be more productive, the resulting produce will be of lower nutritional value."

Robert Mendelsohn, an environmental economist at Yale University, says, "There is no doubt but that quality matters. If scientists can demonstrate a distinct loss of quality, this would be important and could change our impression of the global impact of climate change on agriculture from benign to harmful."

In the study conducted by Peter S. Curtis and his colleagues, they relied on data collected from 159 studies over the past 20 years involving 79 different plant species. Each plant had been subjected to twice the normal levels of CO_2—an amount equal to what is predicted the Earth's atmosphere will contain by 2100. The collective sampling included a diverse number of species such as corn, wheat, rice, blueberries, cranberries, cotton, pasture grasses, and several versions of wild plants.

One of the key concepts Curtis wanted to focus on was the overall reproductive traits of plants such as the number of seeds and fruit, their size, and their nutritional quality because researchers had already confirmed that increasing CO_2 levels did increase the growth of the plant's leaves, stems, and roots.

Curtis stressed the importance of this because the "reproductive traits are key characteristics for predicting the response of communities and ecosystems to global change." In other words, if plants could grow faster but not successfully reproduce, the point was moot.

The results of the study showed that with a doubling of CO_2, total seed weight increased 25 percent, the number of flowers increased 19 percent, individual seed weight increased 4 percent, and the number of seeds increased 16 percent. The study did have some unexpected results, however. According to Curtis, "The surprise is that nitrogen levels actually go down with elevated CO_2 levels, which reduces the nutritional value of these foods because it lowers the protein content. In some cases, nitrogen levels were 15 to 20 percent lower."

Irakli Loladze, an ecologist at Princeton University, says, "The increase in crop productivity does not make up for the fall in nutritional value of the crops—plants today provide 84 percent of the calories people eat worldwide and are also the source of major essential nutrients." He also noticed that levels of micronutrients (such as iron, zinc, and iodine) also dropped as CO_2 levels rose.

Another discovery was that not all crops experienced the same type of growth. Roughly 30 percent had a notable increase in growth—most of these were traditional crops. Wild species, however, used much of their growth energy to produce chemical deterrents and thicker, tougher leaves for protection.

Peter Curtis explained the result: "Six thousand to 10,000 years of breeding have caused domestic plants to put all their energy into seeds and fruits. This leaves them very poorly equipped to deal with the vagaries of nature. But that's okay because we protect them from insects, browsing animals, and disease. Wild plants do not have this luxury."

As CO_2 levels increase, researchers are not sure at this time how the wild plant species will be affected—some will be able to adapt better than others. According to Curtis: "Concerning global warming and this dilemma, there will be winners and losers. We may end up with a bunch of fast growing weedy species."

A study conducted by Ramakrishna Nemani at the University of Montana marked the first vegetation inventory and the effects of global warming as it relates to temperature and precipitation from a global perspective. The study, conducted June 2003, showed that global vegetation had increased 6 percent from 1982 to 1999. In the Amazon region, Nemani says a major cause of vegetation growth is due to less cloud cover, resulting in increased solar radiation. The Amazon region accounts for 42 percent of the global increase.

The atmospheric scientist Dave Schimel, who works at the National Center for Atmospheric Research in Boulder, Colorado, keyed in on the impacts due to absence of cloud cover and its significance for increased vegetation. According to Schimel, "Most studies of the effects of climate change have addressed temperature effects, some have also addressed water effects. But some of the most robust observed changes in climate have been in cloudiness and almost no studies have examined trends in solar radiation. So this is a really interesting new perspective."

The increased vegetation growth has been attributed to different causes in different portions of the world. In the Amazon, the cause has been increased sunlight along with no decrease in rainfall. In areas such as Australia, southern Africa, and India, wetter weather trends over the past few decades are attributed to increased rainfall, such as that received during the monsoon periods.

Negative Effects and Challenges

There are also certain negative effects connected with agricultural practices associated with global warming. Attention is being focused on the gases emitted from cow herds in agricultural areas. According to a report on *LiveScience* in July 2005, eight pregnant Holstein cows are being used in a controlled environment to detect resultant gas levels and determine their negative effects. In fact, researcher Frank Mitloehner of the University of California has eight cows confined inside what he refers to as a "bio-bubble" (a big white tent) to answer an extremely important question: How much gas does a cow actually emit?

This study is unique because it will not only answer the basic question as to how much gas is contributed to the atmosphere by cattle; it has a secondary aspect, as well. Findings will also be used to help write California's first air-quality regulations to involve the agricultural industry—the first attempt of its kind in the nation. As a result, dairies may eventually have to apply for air-quality permits and begin using equipment designed to minimize and eliminate air pollution, just as other industries must currently do.

Based on outdated information (collected in the 1930s), cows produce 12.8 pounds of volatile organic compounds (VOCs) a year. Current estimates of VOCs, however, have been suggested to be as high as 20.6 pounds per cow per year. In order to gain the most precise readings possible, the air is monitored in the confined area by equipment so sensitive that it is able to detect one molecule out of a trillion.

According to Mitloehner, one of the most surprising results of the experiment was that the most significant greenhouse gas emissions were not a result of the manure. According to him, "We thought it was the waste that would lead to the majority of the [greenhouse gas] emissions, but it seemed to have been the animals."

The major contributor to greenhouse gases was the ruminating process. During a cow's digestion process, as food enters the stomach, it mixes with bacteria, which breaks the food down and produces methane (a greenhouse gas). Roughly 20 minutes later, the food returns to the cow's mouth as cud, which it chews, thereby releasing methane into the air. In addition to methane, it also releases methanol and ethanol, as well as VOCs.

As the population becomes more environmentally aware, it has spurred the notion that the agricultural sector may need to be regulated for pollution control as much as any other industrial sector that releases greenhouse gases into the atmosphere. Other researchers, such as J. P. Cativiela of Dairy CARES, an industry-funded environmental group, have countered the notion, however, suggesting that changing a cow's food type may be more effective than investing in and implementing expensive technologies. Others, principally environmentalists, argue that the whole issue is overexaggerated and not as serious as it has been portrayed. For example, Brent Newell of the Center on Race, Poverty, and the Environment says: "It doesn't take into account the lagoons that store the waste or the decomposing feed, the decomposing corn stored in a dairy."

Another negative impact concerns manure management. The decomposition of animal waste in an anaerobic (oxygen-free) environment produces methane. According to the Environmental Protection Agency (EPA), manure storage and treatment systems are responsible for about 9 percent of the total methane emissions in the United States, and 31 percent of the methane emissions coming from the agricultural sector alone.

Methane has an enormous greenhouse gas potential. It is about 21 times more effective at trapping heat than CO_2. In fact, 10 percent of the warming in the United States caused by global warming is due to methane alone. More than 80 percent of the methane emissions that originate from animal wastes come from liquid-based manure management systems that include the following:

- anaerobic lagoons,
- holding tanks,
- manure ponds.

Another management option—that of spreading manure on agricultural fields as fertilizer—does not produce significant amounts of greenhouse gas. Its drawback, however, is that it can be washed off the surface of the ground and infiltrate streams and groundwater, polluting those valuable resources.

According to the EPA, emissions from liquid manure management practices that cause greenhouse gas emissions have been steadily increasing since 1990, due to increasing animal populations. This trend

As urban areas expand, farmlands are being encroached upon. *(USDA)*

is expected to continue. Two major health issues have been identified as a result: odor control and water-quality issues. Also added to the concern is that urban areas are beginning to expand and encroach on rural areas, which means negative impacts for residential areas.

According to the EPA, one solution that has been suggested to combat this problem is the use of anaerobic digester systems. An anaerobic digester is a container similar to a covered lagoon that is designed to hold decomposing manure under warm, oxygen-free conditions that promote the growth of naturally occurring bacteria. The purpose of the bacteria is to digest the manure, producing methane and an effluent that farmers can use in place of untreated manure.

This technology is not new; it has actually existed since the 1920s. During World War II in Europe, when petroleum was scarce, anaerobic digesters used heat and bacteria to eat animal waste, which allowed the Europeans to capture and burn the methane gas for energy. Only during the past decade, this technology has become recognized in the United States as a viable option for the production of energy and mitigation of global warming.

The by-product (methane) produced by these digesters is called "biogas," and it can now be used effectively as an energy source. The gas is collected in perforated pipes and transmitted to an electric generator or boiler. Once the biogas is produced, farmers can use it as an energy source to produce electricity, heat their water, or use it to generate refrigeration. Depending on how much they generate, they may also have the option to sell "extra" electricity directly to local utility companies. Secondary benefits include a reduction of odors, methane emissions, and surface/groundwater contamination. Farmers can also sell the resulting high-quality fertilizer. These related business opportunities help the overall position of U.S. agribusiness.

One example of an area whose agricultural business sector has benefited is Craven Farms of Cloverdale, Oregon. According to a report issued by the EPA, Craven Farms annually produces $24,000 of electricity and $30,000 of digested solids with its biogas system. The EPA contends that "maximizing farm resources in this way may help farmers remain competitive and environmentally sustainable in today's livestock industry."

Also, based on EPA findings, digested wastes are biologically stable when compared with untreated manure, and when they mineralize, they have a higher ammonium content that allows crops to utilize it. They also reduce the likelihood of surface or groundwater pollution. In addition, they reduce pathogen populations in manure, destroy weed seeds (such as from invasive species), and control unpleasant odors.

The federal AgSTAR Program maintains there are about 3,000 livestock facilities across the United States that could install cost-effective biogas recovery systems, with the potential to recover 470,000 tons (426,000 metric tons) of methane. The AgSTAR Program is a voluntary effort jointly sponsored by the U.S. Environmental Protection Agency (EPA), the U.S. Department of Agriculture (USDA), and the U.S. Department of Energy (DOE). The program encourages the use of methane recovery (biogas) technologies at the confined animal feeding operations that manage manure as liquids or slurries. These technologies reduce methane emissions while achieving other environmental benefits.

Fertilizer management is another area of concern. Based on information collected by the EPA, the use of synthetic *nitrogen* and organic fertilizers has contributed to about 36 percent of total U.S. nitrous oxide (N_2O) emissions. Nitrous oxide is about 310 times more effective at trapping heat in the atmosphere than CO_2. While most N_2O is produced naturally through microbial processes in the soil, it is the addition of synthetic N_2O introduced through fertilizers that has led to an increase in emissions from agricultural lands. According to the IPCC, if fertilizer applications are doubled, emissions of N_2O will double. Therefore, the IPCC has recommended that nitrogen fertilizer be used more efficiently in order to reduce N_2O emissions. Simultaneously, reducing the amount of fertilizer used will also decrease the CO_2 emissions that occur during the fertilizer manufacturing process (which relies on energy generated from fossil fuels).

The federal government is also playing a role in an effort to encourage a reduction in the generation of greenhouse gases. Under the Climate Change Action Plan, the U.S. Department of Agriculture is sponsoring projects and demonstrations to help farmers improve their efficiency in the use of nitrogen fertilizers.

The federal Environmental Quality Incentives Program (EQIP), established under the 1996 Farm Bill, offers cost-sharing and incentive payments for land management practices such as nutrient and manure management.

In another program through the USDA's Natural Resource Conservation Service, private insurers in some states now offer crop insurance that protects farmers against potential yield reductions associated with the use of alternative nitrogen management techniques. This insurance provides a "safety net" to farmers so that they will be willing to experiment with other techniques.

According to the Food and Agriculture Organization (FAO) of the United Nations, agriculture and deforestation account for about one-third of global greenhouse gas emissions from human activities. Of these, approximately 80 percent of the total emissions come from developing countries. In addition to the production of crops, livestock in developing countries is also in serious trouble as tens of thousands of animals die from heat waves and lack of water and proper nutrition. In fact, many breeds of livestock cannot be genetically improved fast enough to adapt to climate change.

Based on data from the FAO, a rise in temperature of 5°F (3°C) could jeopardize water supplies for 155 to 600 million people in the Near East alone. Hunger is already a serious problem in many parts of the world, such as Africa. According to the FAO, agriculture could be made nearly carbon neutral using available technologies and proper land management practices.

ACRES OF OPPORTUNITY—GREENHOUSE GAS MITIGATION AND BIOFUELS

Although the agricultural sector is a significant source of greenhouse gases and contributes to the problem, it does have the potential to play a significant role in the overall solution of the problem. The agricultural industry is in a unique position to sequester greenhouse gases, thereby offsetting not only its own emissions but also those of other industries.

According to research done by the Environmental Defense Fund, because the agricultural industry has the ability to offset and counterbalance global warming gases, American agriculture has unique opportunities unlike that available to any other industry. The research also contends that if farmers and livestock owners voluntarily choose to alter some of their on-farm practices, they are also in a unique position to access a new carbon market while simultaneously improving the health of their land and water resources.

According to a report that appeared in *Farm Journal* on January 11, 2006, "Climate changes affect every aspect of how you farm and what you produce."

Based on results from research by NOAA and NASA, farmers know that global warming will probably cause severe droughts in the United States Corn Belt, and will likely increase the occurrence of pests and disease everywhere. While it is true that some areas may experience longer growing seasons that farmers will be able to take advantage of—such as being able to produce two crops instead of just one—they are also aware that long-term negative impacts from climate change will most likely far outweigh any short-term benefits.

Of several opportunities for farmers to contribute to the reduction of global warming by reducing heat-trapping emissions, two major ways where they can make this happen are: (1) increasing the efficiency and storage of carbon in the soil, and (2) producing *renewable* energy. In order to promote the occurrence of these options, the U.S. federal government has put economic incentives in place to encourage farmers to participate in related programs. For instance, farmers who adopt new practices are allowed to enter new markets and be paid for not only what crops they grow but also for the methods by which they grow them.

Mitigation and Sequestration

Farmlands are currently viewed as opportunistic areas of carbon storage if they are managed properly. It is known that the tillage method used is a critical factor in determining the amount of carbon left in the soil. The more the soil is disturbed, the more carbon is released into the atmosphere. Therefore, the goal is to reduce soil disturbance by using no-till or low-till methods instead of traditional till methods. Low-till methods are called "carbon sequestration" methods because they retain carbon in the soil instead of releasing it into the atmosphere. Conserving carbon in the soil not only benefits the environment; it also benefits the farmer. It makes the soil more productive, reduces erosion, and benefits wildlife habitat by making the habitat more fertile.

According to the Pew Center on Global Climate Change, agricultural mitigation and sequestration offers a unique means of managing and reducing greenhouse gas emissions. The studies have identified several opportunities where soil carbon can be increased. One major method involves managing land more efficiently by choosing tillage methods that increase plant residues within the soil. Studies have also determined that if half of the original historic croplands of the United States could be regained, that tens to hundreds of millions of tons of carbon could be stored in the soils annually over the next several decades. In addition to increasing plant residues, slowing the rate of soil organic matter decay would also increase soil carbon reservoirs. Restoration of

Conservation tillage leaves at least 30 percent of the soil covered after planting with last year's crop residue, as shown in this farm in central Iowa. *(Lynn Betts, NRCS)*

No-till planting on the contour in a field in northwest Iowa *(Gene Alexander. NRCS)*

degraded lands can also increase soil carbon. The Pew Center states that these processes of soil sequestration and storage will increase carbon content for about 20 to 30 years, after which time the carbon content stabilizes and cannot store additional carbon.

During farming practices, carbon inputs to soil can be increased by three major methods: (1) using crop rotations with high-residue yields, (2) reducing or eliminating the fallow period between successive crops in annual crop rotations, and (3) by using fertilizer efficiently.

High-residue crops and grasses are those that leave large amounts of plant matter in the soil after the crop is harvested. Good examples of this are corn, sorghum, hay, and pasture grasses, which have significant root systems in the soil.

The elimination of the practice of fallow periods between crops is another method that has proven successful in promoting higher carbon levels in areas such as the western United States. For instance, the practice of planting winter wheat along with summer season crops such as corn, millet, sorghum, and beans utilizing a no-till practice not only increases soil carbon but also improves soil moisture—an important and scarce resource for the American Southwest.

It is also important to apply manure, nitrogen fertilizers, and irrigation efficiently. If too much fertilizer is used, it offsets the benefits of the carbon stored in the soil because it increases the levels of N_2O. Each area needs to be assessed for its specific needs so that a proper balance is reached, and an area gets only the fertilizer it will actually utilize—no more (which will simply be left behind in the soil or washed into another source by the irrigation).

Reducing the impact from tillage is also important. Soil tillage in general accelerates organic matter decomposition by warming up the soil and breaking it apart where it can decay more rapidly, releasing greenhouse gases such as CO_2 into the atmosphere. Traditional tillage methods are the most invasive (such as *moldboard plowing*) because they completely invert (turn over) the soil. Reduced tillage methods are better because they plow to shallower depths, cause much less soil mixing, and keep more of the crop residue within the soil. The best alternative is the no-till practice, in which crops are sown by cutting a narrow slot in the soil for the seed and herbicides. This method results in the least amount of soil disturbance.

Pastures and rangelands can also retain large amounts of soil carbon. Not only do their extensive root systems store high proportions of carbon underground, but they protect the ground surface year round with plant cover, and they also help strengthen the soil, keeping the carbon fixed in place. One way to increase carbon storage on these lands is to increase irrigation, plant legumes, or plant improved grass species in humid areas. In the semi-arid western rangelands, there are fewer options available—usually limited to changing species composition. Restoration of serious degraded lands is also an option to increase carbon storage. *Land-use change* is another method. For instance, land that is currently used for crop production can be converted into hay or pasture lands.

Carbon sequestration rates vary geographically depending on climate, soil types, topography, past management history, the current condition of the land, and the current land management practices. Although they may vary, enough research has been done to date to enable the agricultural industry to make positive steps toward combating global warming.

Production of Biofuels

The Environmental Defense Fund has determined that "a key factor in how effective biofuels are in fighting global warming is the energy efficiency of their production methods. These include everything from running plows and harvesters to manufacturing pesticides and fertilizer to converting the material into fuel and transporting it. Improving land use through sustainable practices such as no-till farming, and boosting energy efficiencies make biofuels more effective at reducing heat-trapping pollution."

Based on research conducted by the Pew Center, dedicated energy crops could supply 6 to 12 percent of the current U.S. energy demand without significantly raising food prices. The resultant biomass could be used for a number of applications: to produce heat, power, and for fuel. Currently, the largest markets are in the generation of steam and electricity from forest biomass and fuel (ethanol) from corn. In the future, there will be the production of dedicated energy crops (such as grasses) and agricultural residues. Research by the Pew Center has indicated that producing these energy sources would require the use of 15 percent of U.S. agricultural lands. The upper limit on cost-effective production of ethanol from cornstarch is estimated at 10 billion gallons (38 billion liters) per year, or less than 4 percent of the U.S. transportation demand. In addition, production of ethanol from agricultural residues and energy crops could provide 57 billion to 84 billion gallons (216 billion–318 billion liters) of transportation fuels annually.

Several types of biomass can currently be produced for three major categories: heat, electricity, and transportation fuel. They are as follows:

- grass biomass—heating homes
- short rotation woody crops—heating homes and transportation
- waste vegetable oil—heating homes and transportation
- straight vegetable oil—heating homes and transportation

- biodiesel—transportation
- ethanol—transportation
- corn grain—heating homes

There are advantages to farmers for producing biofuels. As energy prices rise, farms that reduce their energy use increase efficiency. Those that produce fuels will be more financially stable in the country's currently unstable fossil fuels market. In addition, producing and selling biofuels provides additional income to farmers. Biofuel production also supports and strengthens the local economy rather than foreign economies.

AGRICULTURE AND FOOD SUPPLY

Climate change plays an important role in farm productivity, and while some areas may benefit from having longer growing seasons, climate change will also bring an increased potential for heat waves, droughts, and floods. These conditions will pose challenges for farmers and the production of food, especially in regions where water supplies and soil moisture conditions become degraded.

According to the IPCC in their most recent report released in 2007: "Recent studies indicate that increased frequency of heat stress, droughts and floods negatively affect crop yields and livestock beyond the impacts of mean climate change, creating the possibility for surprises, with impacts that are larger, and occurring earlier, than predicted using changes in mean variables alone. This is especially the case for subsistence sectors at low latitudes. Climate variability and change also modify the risks of fires, pest and pathogen outbreak, negatively affecting food, fiber and forestry."

The U.S. Environmental Protection Agency (EPA) has identified several climatic factors that directly affect agricultural productivity, as follows:

- change in climatic variability and extreme events,
- average temperature increase,
- change in rainfall amount and patterns,
- atmospheric pollution levels,
- rising atmospheric concentrations of CO_2.

Each of these individual factors in itself can have a significant effect on agriculture. According to the EPA, climate studies, when related to agricultural impacts, so far have only considered one or two of the above-mentioned climate change factors. The EPA has currently identified the need for more work to be done in the analysis and modeling of the simultaneous influence of the full set of listed factors on agricultural production. In that way, a more realistic view of the effect of global warming on vegetation and agriculture can be reached.

Change in Climatic Variability and Extreme Events

This aspect concerns small-scale significant changes that can occur at local levels, such as the devastating occurrence of heat waves, drought, hurricanes, and floods. One of the greatest limitations of current models is being able to predict events at small, or local, scales—models are best at predictions of large-scale trends and occurrences. Therefore, extreme events that tend to involve limited areas (such as heat waves, floods, etc.), will also involve limited areas of agriculture, making future predictions as to the magnitude difficult. Extreme events could devastate and destroy all agricultural areas within them.

Average Temperature Increase

If there is an increase in average temperature, several things could happen. One positive effect is that it could lengthen the growing season in areas that currently have a short growing season. In areas that currently already get too arid and hot during the summer, an average temperature increase would not be welcome; everything would dry out and die. In addition, it would increase soil evaporation rates, keeping the soil from holding the water necessary to support plant life. Once the soil becomes stressed enough, the effects of drought will become much more severe, possibly leading to *desertification.*

Change in Rainfall Amounts and Patterns

If rainfall patterns change, soil moisture content and erosion rates can change. If this happens, crop yields will be reduced. According to the IPCC, the subtropical regions will experience the most pronounced decrease in precipitation—they believe some areas may experience up to 20 percent less precipitation than usual. As evidenced by an increase in flooding and hurricane events, the number of extreme precipitation events is also expected to increase.

Atmospheric Pollution Levels

As pollution levels increase with the use of fossil fuels (emissions from industrial smokestacks and exhaust from cars), ozone concentrations at the Earth's surface increase. Increased concentrations are unhealthy for the growth of plants and may limit the growth of crops. Because ozone will be working against the growth of vegetation, the IPCC has postulated that any increase in CO_2 that may initially benefit the growth of crops will be negatively offset by the emission of ozone.

Rising Atmospheric Concentrations of CO_2

It is expected that an initial increase in CO_2 will actually behave like a fertilizer and enhance the growth of crops such as rice, wheat, and soybeans. The IPCC cautions, however, that even though it may initially increase production, there will be a slowing that occurs as temperature and precipitation changes may counteract the "beneficial CO_2" effects.

When looking strictly at North America, research by the IPCC in their 2007 report concluded: "Moderate climate change will probably increase the production of 'rain fed' agriculture." Estimates of the increases, however, have been lowered from what they once were. Over the next few decades, agricultural yields are expected to increase by 5 to 20 percent. Food production is expected to benefit some areas of the continent but decrease in others. The U.S. Great Plains/Canadian prairies regions are expected to be some of the most vulnerable to climate change. They have also determined that the crops grown closest to climate thresholds (regions of abrupt change) such as California's wine grapes will suffer the most serious decreases in yields and quality. Crop production in the Southern Plains, Appalachia, the Corn Belt, and in the Delta States are expected to have a decrease in productivity of 16 to 21 percent. On the other hand, fruit crops in northern regions such as the Great Lakes and eastern Canada are expected to increase because the length of the growing season and temperatures will increase. Northern Europe is expected to produce an increase in cereal crops. The low-latitude (tropical) countries are expected to have a decrease in grain production.

Overall, vulnerability is also economic in nature. The agriculture in developed countries—such as the United States, western European countries, Canada, and other industrialized countries—is expected to be less vulnerable than the agriculture of developing nations, such as countries in the Tropics. In the poorer nations, problems arise because farmers have limited opportunities to adapt because they lack the land, equipment, capital, and technology.

Land management practices are a critical issue when farmers are faced with global warming. By varying which crops are grown from season to season and using new technology, farmers can better adjust appropriately to climate change. As food becomes scarcer and demands are put on the land, different areas of the world will experience different degrees of hunger. Those that will be affected the most are in the developing countries of Africa, Asia, and South America.

The IPCC has also warned that impacts to agriculture will become more severe as climate change continues. If CO_2 concentrations rise beyond double what they are today, then damages may become a significant loss to the U.S. agricultural economy. In other words, as CO_2 levels continue to rise, even areas that experienced a temporary increase in yield will suddenly face a decrease in yield. In addition, there will also be a northern shift in invasive weeds.

Disease pressures on crops and domestic animals are also expected to increase with earlier springs and warmer winters, which will allow rapid reproduction and higher survival rates of pathogens and parasites. The regional distribution of warming and changes in rainfall will also affect the spatial and temporal distribution of disease. According to the U.S. Climate Change Science Program, the following list defines their findings related to rangelands and pasturelands:

- Projected increases in temperature and a lengthening of the growing season will likely extend forage production into late fall and early spring, thereby decreasing the need for winter-season forage reserves.

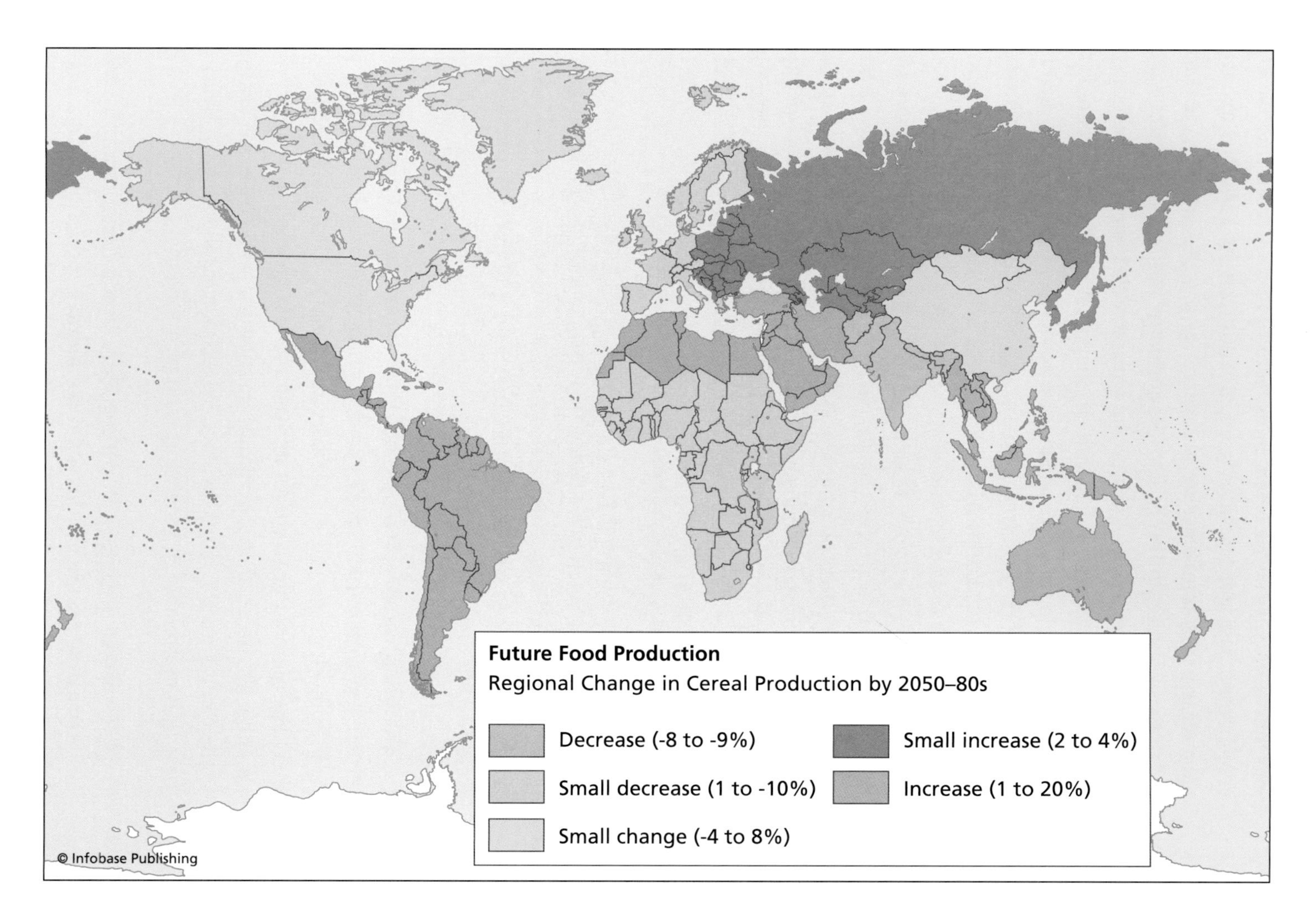

Future Food Production
Regional Change in Cereal Production by 2050–80s
Decrease (-8 to -9%)
Small increase (2 to 4%)
Small decrease (1 to -10%)
Increase (1 to 20%)
Small change (-4 to 8%)
© Infobase Publishing

- In both pasture and rangelands, shifts in optimal temperatures for photosynthesis might be expected under elevated CO_2.
- Climate change–induced shifts in plant species are already under way in rangelands. Establishment of perennial herbaceous species is reducing soil water availability early in the growing season.
- Increasing CO_2 will alter forage quality; in nitrogen-limited native rangeland systems, CO_2-induced reduction in nitrogen and increase in fibrous plants (woody shrubs, trees) may lower quality.

The following list pertains to their findings relating to animal management:

- Higher temperatures will very likely reduce livestock production during the summer season.
- For ruminants (cattle), current management systems generally do not provide shelter to buffer the adverse effects of changing climate; such protection is more frequently available for nonruminants such as swine or poultry.
- Benefits from extended forage season production and reduced need for winter-season forage reserves will very likely have significant impact on livestock operations.
- Shifts in rangeland and pastureland plant productivity and type will likely have a significant impact on livestock operations.

According to the Pew Center on Global Climate Change, the added stress of global warming could add enough stress to threaten local or regional food supplies. In addition, population growth, economic hardship, poor land management practices, and political instability can impair a nation's ability to feed its population and cope with global warming. The IPCC estimates that even without climate change, the number of

(opposite page) Climate change will threaten food security worldwide with the largest impact to developing countries.

people at risk of hunger and malnourishment will increase from 500 million currently to more than 640 million by 2060. Global warming will increase those numbers. In terms of food security, sub-Saharan Africa is at most risk for food scarcity. Lacking rainfall and having a high level of evapotranspiration, more than half of its land is irrigated. It is also very low in organic matter, making it highly nonfertile.

According to a report in *Discovery News,* farmers hoping to use desalinated water as a resource in water-scarce areas to irrigate crops have had to overcome a hurdle. At a desalination plant currently operated in Israel, it was determined that the converted water was lacking in calcium and magnesium, but was overloaded with boron. While not a problem for drinking water, it was for the production of crops—but with the addition of fertilizer and some other adjustments, the water could be treated so that it could be safely used for irrigation.

Farmers also discovered that using desalinated water also helps reduce the existing salt content in soils—particularly those soils found in the arid and semi-arid regions of the world. While desalinated water was not originally intended to be used as irrigation water, the concept is slowly gaining ground, currently being used by Australia, Israel, and Spain. Currently, 69 percent of the global water supply is used for irrigation.

According to Andrew Chang, director of the Center for Water Resources at the University of California, "Desalinated water hasn't yet made it into agricultural use in the United States, but that's not to say it won't someday."

Currently, a municipal desalination plant in Santa Barbara, California, is functioning as a reserve against drought. As global warming intensifies, desalination may become a key factor in providing a critical resource for both drinking and agricultural water. As technology becomes more sophisticated, scientists will continue to work on better ways to manage global warming issues.

Because the agricultural sector is critical to everyone on Earth, it is especially important to understand the cause-and-effect relationship between it, greenhouse gases, and global warming. Public education is especially important in order to be assured of maintaining an adequate global food supply.

4

Deforestation and Greenhouse Gases

The world's forests also play a critical role in climate change through global warming. Currently, the trees in forests act as a reservoir, storing huge amounts of carbon, sequestering it from the atmosphere. The problem arises, however, when the trees are cut and burned. This process takes the carbon from the "sink" it has been stored in and releases enormous amounts of CO_2 into the atmosphere. Currently, huge areas of forests are being cut and burned at alarming rates. This uncontrolled deforestation is not only releasing CO_2 into the atmosphere, but also changing the climate regime worldwide. This chapter discusses the pertinent issues of deforestation, the impacts of climate change, and the forest resources currently at risk as global warming escalates.

FORESTS' ROLE IN CLIMATE CHANGE

Forests represent a critical resource on Earth because they provide habitat to more than half of the Earth's living species, as well as help slow global warming through the storage and sequestration of carbon. Forests also affect the rainfall regime of an area and are critical sources of food, drinking water, and medicine.

Through the sequestration of carbon, forests have a direct effect on the climate. The vegetation and soils in the forest help drive the global carbon cycle by sequestering CO_2 through photosynthesis and then releasing it through respiration. All forests sequester carbon, but as forests age and the trees become more mature, their storage capacity decreases.

A major conservation issue facing the world today is deforestation—the rapid clearing and destruction of many areas of forest. Forested areas—particularly rain forest areas—are currently being cleared for other activities such as agriculture and rangeland for grazing livestock. Many of these areas are also being logged for their precious hardwood resources such as teak, mahogany, merbau, melina, and okoume. Mining operations claim yet other forested areas, and others are lost to human-caused wildfires.

As soon as the forest is destroyed, the carbon that was sequestered for long periods of time in the trees and soil is released to the atmosphere, forming CO_2. In fact, according to the Union of Concerned Scientists (UCS), tropical deforestation accounts for about 20 percent of the total human-caused CO_2 emissions every year.

Currently, forests in the United States are considered to be net carbon "sinks" because they sequester (store) more carbon than they release. One reason for this is that many of the lands in the Northeast that were once cleared of forests are now being reclaimed by new forest growth—the more forest growth there is, the more net storage of carbon there is. Other factors are also helping give new forests a boost, such as better control of wildfire incidences, changes in the way trees are harvested for wood products, and increased growth in forests due to increased levels of CO_2 in the atmosphere in recent years. Even though U.S. forests have shown an increase in carbon sequestration since the 1980s, the UCS warns that the trend will most likely slow down as physical processes change or evolve. They have also stated that both forest and land-use practices, if managed correctly, have the potential to reduce net carbon emissions 10 to 20 percent of the projected fossil fuel emissions through 2050. In the United States alone, it may be possible to sequester 44 to 88 million tons (40 to 80 million metric tons) of carbon each year—an equivalent of 3 to 5 percent of current annual U.S. fossil fuel emissions. When looking at forests worldwide, there is even greater carbon sequestration potential by forests in the tropical and subtropical regions.

When looking for ways to manage, or mitigate, the effects of global warming, proper management of forested areas in order to protect, restore, and sustain the resources in forests is one of the more promising methods of offsetting the negative effects of global warming. The UCS has identified three different methods of forest-based mitigation:

- Conservation of existing forests. By leaving forests uncut, it preserves the carbon already stored in the trees, utilizing it as a carbon sink.
- Sequestration by increasing forest carbon absorption capacity. This can occur from planting new trees or through the natural regeneration of the forest on its own. Forests must be managed properly for this to occur.
- Substitution of sustainably produced biological products. This method involves substituting wood products for materials requiring energy-intensive production (such as aluminum or concrete), and substituting woody biomass for fossil fuels as an energy source.

The UCS also points out that if forests are managed properly in order to combat global warming, there are other resulting benefits as well. Benefits can be environmental, such as protecting *biodiversity* and watersheds; supporting food cycles; promoting healthy ecosystems, providing for humans; and conserving wildlife, vegetation, and water resources. Benefits can also be social, such as providing new jobs in local businesses.

One of the major problems in controlling deforestation is the lack of education. Much of those forested areas are cut for economic reasons. Two of the principal reasons are harvesting the valuable hardwoods strictly for their international market value and removing the forest in order to convert the land to farming and/or rangeland. When a promised income from these activities is presented to poor landowners in the Tropics, they are often accepted in order to gain much needed income to support starving families. What needs to be put in place are financial incentives to forest landowners to preserve the forests' long-term goods and services, outweighing any short-term gains that could be derived from forest destruction. To date, this has not happened, but scientists and resource specialists such as those with the U.S. Natural Resource Conservation Service (NRCS), as well as private interest groups such as the UCS, believe that if there were market-based incentives to promote forest conservation that were more enticing than those to destroy the forests, much of the problem could be solved, thereby decreasing the negative effects of global warming.

One of the proposed solutions to this problem includes increasing forest area and size, which improves biodiversity and forest ecosystem dynamics. The UCS has identified the following set of market-based approaches that they believe would promote forest-based mitigation of climate change:

- Slow deforestation internationally through the Clean Development Mechanism (CDM) and other international investments in forest conservation.
- Create a carbon market that recognizes domestic forest carbon values and creates strong incentives for reducing emissions in the United States by protecting and restoring natural forests.
- Manage timber production forests for carbon and other environmental values.
- Preserve the integrity of mature forests when managing for timber or biomass.
- Maintain historical fire regimes.
- Maintain environmental safeguards on U.S. public forested lands.

The Clean Development Mechanism (CDM) is part of the Kyoto Protocol agreement (an international agreement designed to control global warming), which allows industrialized countries to invest in various emission reduction projects that are taking place in developing countries, where emissions abatement is economically efficient. The CDM enables developed countries to apply certified emission reductions to these areas and slow deforestation.

In Bonn, Germany, at the July 2001 climate policy negotiations of the UN Framework Convention on Climate Change (UNFCCC), it was voted to grant CDM credits to projects that grow trees (afforestation and reforestation) in developing countries but not those that protect existing forests from being cleared or degraded. This decision applies only to the Kyoto Protocol's "first commitment period" in effect from 2008 to 2012.

While the CDM helps in areas of afforestation and reforestation, it severely lacks in probably the most important aspect: that of protecting and conserving the threatened natural forests. Conservation of existing forests is critical in order to slow CO_2 emissions and prevent global warming, making the CDM one avenue that needs to be pursued in order to change the policy and add a clause to protect threatened natural forests.

In response to this, the U.S. Congress is currently considering legislation that will give U.S. businesses tax credits to invest in international forest-based projects that are designed to reduce global warming and protect biodiversity. A draft bill is now being considered by the House Energy and Commerce Committee that will provide provisions for protecting tropical forests in cap-and-trade legislation. The business credits from it would generate about $12–$15 billion by 2015 for international forest conservation. This draft has received the support of several key American businesses, such as American Electric Power, Conservation International, Duke Energy, Environmental Defense Fund, El Paso Corporation, National Wildlife Federation, Marriott International, Mercy Corps, Natural Resources Defense Council, PG&E Corporation, Sierra Club, The Nature Conservancy, Union of Concerned Scientists, The Walt Disney Company, Wildlife Conservation Society, and the Woods Hole Research Center. It would allow companies to receive credit for reducing emissions by financing activities that protect forests in tropical countries. It also calls for 5 percent of proceeds from the auctioning of greenhouse gas emissions allowances under a cap-and-trade system to go towards funding forest conservation projects.

The initiative is attractive to U.S. business because if offers a lower cost method for complying with emission caps. Environmental groups see the concept as a way to potentially generate substantial funds for forest conservation and sustainable development, while also helping mitigate climate change by reducing greenhouse gas emissions from deforestation.

The UCS believes the U.S. federal government needs to play an important role in controlling global warming. They believe that whether the U.S. government ever ratifies the Kyoto Protocol or not (which it has not yet), that it should still put mandatory limits (called "caps") on carbon emissions and also create an economy-wide domestic carbon market equivalent to what is stipulated in the protocol. Referred to as a "cap-and-trade" system, it allows the market flexibility to find the most efficient and innovative ways of meeting their mandatory emission limits. They believe that if businesses voluntarily take action to reduce carbon emissions, they should be rewarded by tax credits or other government incentives. There are several legislative proposals currently under consideration to provide such incentives for voluntary CO_2 sequestration or emission reduction projects on private lands. Such voluntary activities include reforestation and changes in agricultural practices that lead to increased carbon storage in the soil, such as low- or no-till plowing methods. Rewards for these efforts could include tax credits or subsidies to land owners.

Forests that are currently utilized for timber production also need to be managed for carbon emissions. There are several methods by which this can be achieved:

- Expanding forest area by allowing the regrowth of trees, enabling the forest to increase in area naturally.
- Allowing trees to grow larger (which increases carbon storage).
- Establishing conservation set-asides (no-harvest areas) within productive forests.
- Using harvesting methods that reduce damage and waste.

When these conservation measures are followed, the benefits are twofold. Not only do they allow greater carbon storage, they also improve ecosystem health, which makes them more resilient in the face of global warming. When they have greater stability, they are better able to withstand the disturbances associated with climate change.

Controlling the harvesting dates within forests is also critical. Young trees grow more rapidly than mature trees, but as trees age, the more carbon they are able to sequester. The problem arises, however, when timber companies have a strong economic incentive to harvest at a faster rate (more trees, more revenue). According to L. A. Wayburn, et al., lengthening the time between harvests could significantly increase carbon storage in the Pacific Northwest and southeastern United States.

According to R. F. Noss of the University of Central Florida, "Establishing a carbon market and a sound regulatory framework could provide financial incentives to lengthen harvest cycles. Reducing damage to non-harvested trees and disturbance of forest soils during logging operations can also substantially reduce CO_2 emissions."

Another major issue of concern is the widespread, but inaccurate, belief that logging and clearing mature forests and replacing them with fast-growing younger trees will benefit the climate by sequestering CO_2. While it is true that young trees grow and sequester carbon quickly, the disturbance of stored carbon when mature forests are logged is also important. When a forest is logged, some of its carbon is stored for long periods of time (decades or longer) in wood products. But what may not be considered is that large quantities of CO_2 are also released to the atmosphere as soon as the forest is disturbed. This happens through the disruption of the soil and the carbon stored within it, as well as decomposition of the leaves, branches, and other biomass left behind over time. According to M. E. Harmon, et al., one study found that even when storage of carbon in timber products is considered, the conversion of 12 million acres (5 million hectares) of mature forests to plantations in the Pacific Northwest over the last 100 years resulted in a net increase of more than 1.5 billion tons (1.4 billion metric tons) of carbon to the atmosphere.

Forest managers also warn that historical forest fire regimes should be left in their natural cycles; they should not be changed in order to increase carbon storage. Forest fires release significant amounts of CO_2 into the atmosphere. They also contribute up to 20 percent of the annual global emissions of methane and nitrogen oxide, which are both potent greenhouse gases. Wildfires (not anthropogenically induced) are natural occurrences, however, and forest managers such as those with the U.S. Forest Service believe that natural fires should not be purposely suppressed. Although not the case historically, they now believe that most forests develop in balance with a natural fire regime that is critical to ecosystem health. Some species of pine, for example, need infrequent hot fires for seed germination. The danger in suppression is that it leads to excess fuel accumulation (extra biomass, including dead vegetation). This exacerbates the problem because once a wildfire does start it will be more catastrophic and release even larger amounts of carbon into the atmosphere.

In the United States, the federal government has a major responsibility for land management because 42 percent of all U.S. forests, including most of the old growth forests, are on public lands managed by the Forest Service, giving them a key role in helping control activities that affect global warming.

CARBON SINKS

According to the World Resources Institute, the world's forests are natural regulators of CO_2 in the atmosphere. While forests are effective sinks for CO_2, if they are being cut down, then they cannot be counted on in the future to absorb the CO_2 being released by fossil fuels and other sources.

According to a study conducted by NASA, scientists estimate that there are roughly 1.1 to 2.2 billion tons (1 to 2 billion metric tons) of carbon generated each year that are "missing" from the global carbon budget. They have determined that 15 to 30 percent of the carbon released into the atmosphere from the burning of fossil fuels is currently unaccounted for in the Earth's carbon budget.

According to Steven C. Wofsy, an environmental scientist at Harvard University, scientists assume that the "missing carbon" is being absorbed by vegetation, and it is important to be able to predict how much CO_2 is generated so that questions such as how much will temperatures rise and how will this affect humans and ecosystems can be answered.

Trees consist of about 50 percent carbon. Immense amounts can be stored in forests, such as these giant redwoods in California. *(NPS)*

Scientists to date have calculated that globally humans release around 7 billion tons of carbon each year. Of that amount, 3 billion tons stay long-term in the atmosphere, and 2 billion tons are absorbed into the oceans. What they do not know is where the rest goes; they assume it is stored long-term in vegetation. And if it is stored in vegetation, are the 2 billion tons stored temporarily or permanently?

In order to answer that question, NASA dispatched an interdisciplinary scientific team called the Boreal Ecosystem-Atmosphere Study (BOREAS) from 1994 to 1997 to cover two Canadian provinces. Teams from 85 nations participated in a quest to find an answer to this question.

CO_2 is measured at different latitudes worldwide, and the data is entered into models that track the circulation of air over a period of time. Through analysis of the models, NASA scientists determined that most of the atmospheric CO_2 was being absorbed somewhere above 40°N latitude in the Northern Hemisphere. NASA scientists also estimate that the boreal forest covers 21 percent of the Earth's forested land surface, contains 13 percent of all the carbon stored in biomass, and holds about 43 percent of all the world's carbon that is stored in soil.

According to Forrest G. Hall, one of the BOREAS project scientists, "For the past 7,000 years, the boreal forest floor has been accumulating carbon at a rate of about 30 grams (roughly one ounce) per square meter per year. When you walk through the Canadian boreal forest, you can literally go from ankle-deep to in over your head in carbon litter. In some places where there were once holes or ravines, we have measured peat several meters deep."

Hall also says that the boreal ecosystem has been taking in an average of 0.66 billion tons (0.6 billion metric tons) of carbon per year for roughly the last 7,000 years. As straightforward as all this sounds, however, it is not. The boreal forest is both a sink (storage area) and a source of carbon because it both absorbs and releases CO_2. According to Hall, on whether the boreal forests could be the net sink of carbon, "We won't know until we add up the contributions from all the kinds of forest that make up the boreal ecosystem. That will require satellite-derived maps of the entire circumpolar boreal region that enable us to sort out the different types of forests and assess their productivity."

What they determined was surprising. From late May through July (spring/summer), the boreal forest "inhaled" 1 to 1.5 grams of carbon per square meter per day. In August and September (the hottest, driest period), carbon intake was almost zero. In October, the forest "exhaled" carbon back into the atmosphere at 0.6 to 0.8 grams per square meter per day.

According to Steven Wofsy, "We found that what's going on in the soils is more important than what is going on in the trees. We found that the boreal peat is drying and thawing out and that the soils will release CO_2 faster than the trees can absorb it."

Hall agrees, stating, "If the boreal soils were to dry, it might increase soil carbon decomposition, but the carbon loss might be offset by the intake of carbon by trees as they grow."

The general trend they observed was that years with warmed springs and falls were better for plant uptake of carbon but resulted in increased rates of carbon loss. It is during these years that boreal forests become a net source rather than a sink. The scientists on the team think that the boreal carbon budget is very sensitive to climate change, and subtle changes in the climate can make the difference between the boreal forest as a carbon source or sink; this difference can show up on a global scale.

This study has major implications for the global climate. Because of boreal forests' climate sensitivity, it responds quickly to changes in climate. In addition, temperature increases are happening the fastest at the high latitude continental interiors—exactly where most of the boreal forests are located. Over the past 40 years, average temperatures have increased by as much as 2.08°F (1.25°C) per decade. Therefore, if the Earth's "missing carbon" is being stored there, and if the recent warming trend is turning the boreal forests into a source of carbon, then there could be an alarming rate of carbon buildup in the atmosphere. In fact, scientists are concerned that by 2100, the amount of CO_2 in the atmosphere might double preindustrial levels—to 560 ppm (in 1990 it was 353 ppm). If this kind of warming happened, it would thaw the frozen boreal soil and permafrost, releasing CO_2 and methane, thereby increasing levels in the atmosphere.

Carbon sinks have been proposed as one way to proactively tackle global warming by "soaking" up excess CO_2. One alternative—planting new forests in deforested areas (reforestation)—is a popular one in locations with huge logging industries and large forests, such as the United States, Canada, some Latin American nations, and Indonesia.

Some environmentalists are critical of simply relying on reforestation as a solution for eliminating large areas of natural forest, however. According to Global Issues Organization, this approach legitimizes continued destruction of old growth and pristine forests that have diverse, rich, well-established ecosystems. The organization also contends that it only provides a "quick fix" but does not directly solve the problem by reducing CO_2 emissions to begin with.

Global Issues also points out that the use of carbon sinks in general is a major "loophole" in the Kyoto Protocol—that if carbon sinks can be counted toward emissions reductions credit, then industrialized countries would be able to meet their commitments while reducing emissions by less than would otherwise be required. Carbon sinks also have the caveat that when vegetation naturally dies or is burned, it releases all the stored carbon to the atmosphere—which turns it quickly from a sink to a source without much warning.

According to Simon Retallack, author of *The Kyoto Loopholes,* "As biologists point out, there is not yet enough data on natural carbon cycling to establish full accounting and verification procedures for carbon sinks. The science simply does not exist to enable prediction of exactly how much carbon is being absorbed by a country's sinks and whether the carbon moving into forests or soils will actually stay there."

IMPACTS OF DEFORESTATION

Deforestation is occurring today at alarming rates, as determined through the analysis of satellite imagery. It accounts for about 20 percent of the heat-trapping gas emissions worldwide. The Food and Agriculture Organization (FAO) estimates current tropical deforestation at 53,000 square miles per year (15,400,000 ha/yr). This equates to an area roughly the size of North Carolina being deforested each year.

Tropical forests hold enormous amounts of carbon. In fact, the plants and soil of tropical forests hold 507 to 634 billion tons (460 to 575 billion metric tons) of carbon worldwide. This equates to each acre of tropical forest storing roughly 198 tons (180 metric tons) of carbon.

Deforestation threatens to upset the global CO_2 balance. Roughly 53,000 square miles (137,269.4 km^2) are lost per year. *(Rhett Butler, Mongabay.com)*

When a forest is cut and replaced by pastures or cropland, the carbon that was stored in the tree trunks joins with the oxygen and is released into the atmosphere as CO_2. Because the wood in a tree is about 50 percent carbon, deforestation has a significant effect on global warming and the global carbon cycle. In fact, according to the Tropical Rainforest Information Center, from the mid-1800s to 1990, worldwide deforestation released 134 billion tons (122 billion metric tons) of carbon into the atmosphere. Currently, 1.8 billion tons (1.6 billion metric tons) is released to the atmosphere each year. As a comparison, all of the fossil fuels (coal, oil, and gas) release about 6 billion tons (5 billion metric tons) per year. Tropical deforestation is the largest source of emissions for many developing countries.

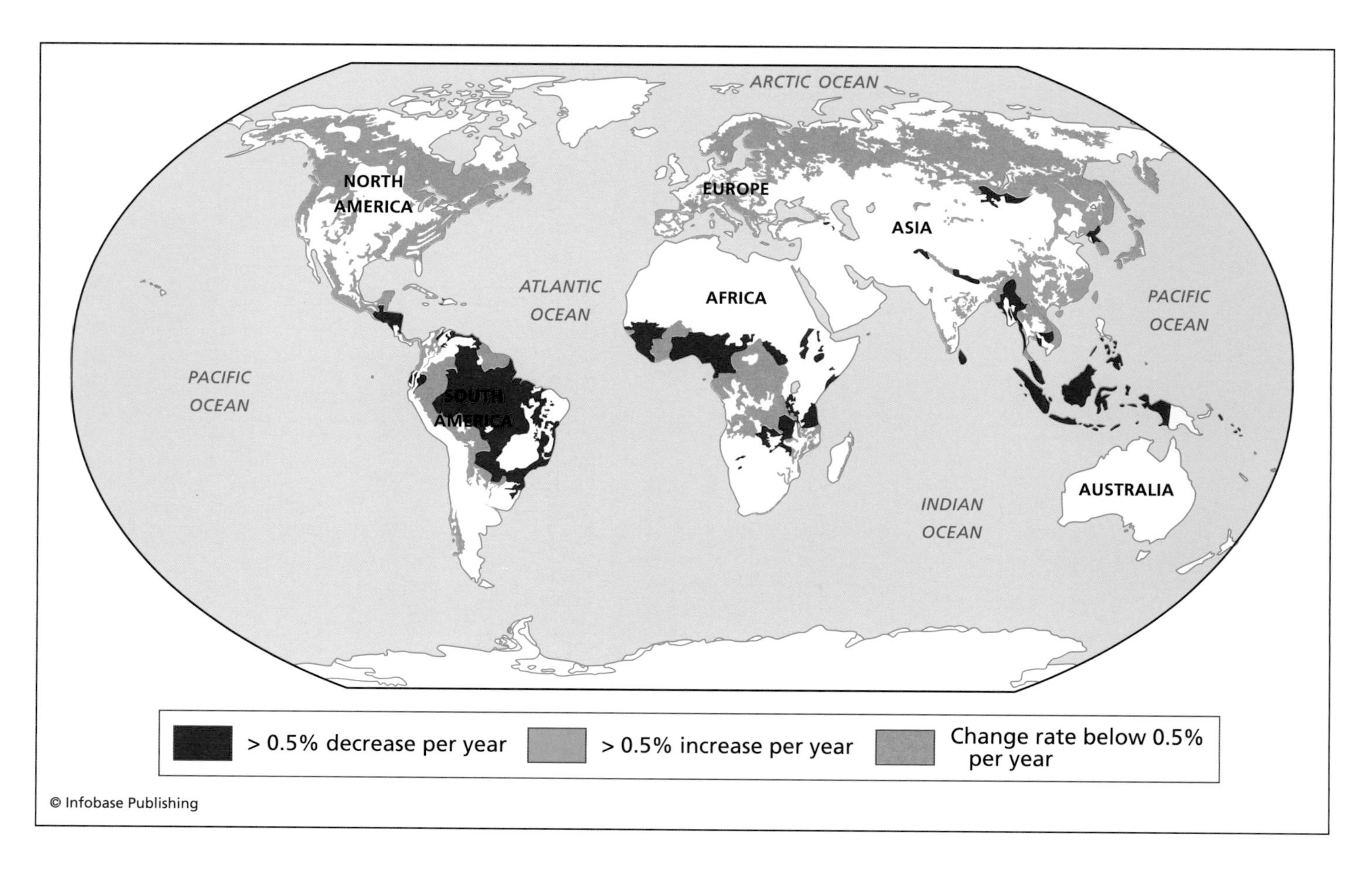
ARCTIC OCEAN
NORTH AMERICA
EUROPE
ASIA
ATLANTIC OCEAN
AFRICA
PACIFIC OCEAN
PACIFIC OCEAN
SOUTH AMERICA
AUSTRALIA
INDIAN OCEAN
> 0.5% decrease per year
> 0.5% increase per year
Change rate below 0.5% per year
© Infobase Publishing

According to Peter Frumhoff of the Union of Concerned Scientists (UCS), he and an international team of 11 research scientists found that if deforestation rates were cut in half by 2050, it would amount to 12 percent of the emissions reductions needed to keep the concentrations of heat-trapping gases in the atmosphere at relatively safe levels.

There are multiple impacts of deforestation. Many of the most severe impacts will be to the tropical rain forests. Even though they cover only about 7 percent of the Earth's land surface, they provide a habitat for about 50 percent of all the known species on Earth. Some of these endemic species (species found only in that particular area) have become so specialized in their respective habitat *niches,* that if the climate changes, which then causes the ecosystem to also change, this will threaten the health and existence of many species, even to the point of extinction. In addition to the species that are destroyed, the ones that remain behind in the isolated remaining forest fragments become very vulnerable and sometimes are threatened to extinction themselves. The outer margins of the remaining forests become dried out as well and are also subjected to die-off.

Two other major impacts in addition to the loss of biodiversity is the loss of natural resources such as timber, fruit, nuts, medicine, oils, resins, latex, and spices; as well as the resultant economic impact along with the extreme reductions of genetic diversity. The loss of genetic diversity could mean a huge loss for the future health of humans. Hidden in the genes of plants, animals, fungi, and bacteria that may not have even been discovered yet could be the cures for diseases such as cancer, diabetes, muscular dystrophy, and Alzheimer's disease. The United Nations FAO states, "The keys to improving the yield and

(opposite page) This represents the net change in forest area between 2000 and 2005. Most of the decrease is occurring in the developing countries, where many people depend solely on income from forests. *(modeled after IPCC)*

nutritional quality of foods may also be found inside the rain forest and it will be crucial for feeding the nearly ten billion people the Earth will likely need to support in the coming decades."

Two of the largest climate impacts will revolve around rainfall and temperature. One-third of the rain that falls in the tropical rain forest is rain that was generated in the water cycle by the rain forest itself. Water is recycled locally as it is evaporated from the soil and vegeta-

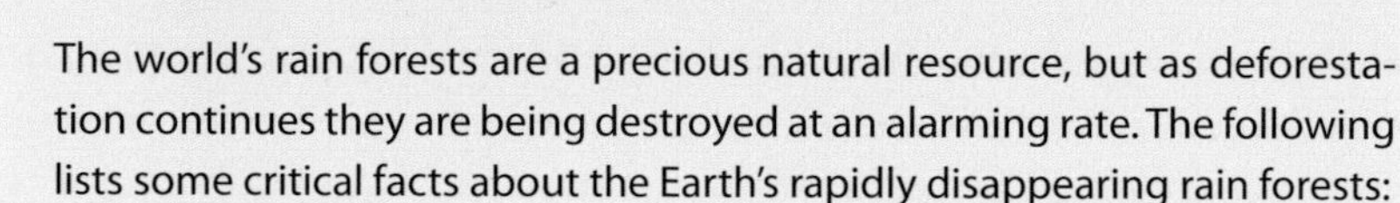

RAIN FOREST FACTS

The world's rain forests are a precious natural resource, but as deforestation continues they are being destroyed at an alarming rate. The following lists some critical facts about the Earth's rapidly disappearing rain forests:

- Almost 50 percent of the Earth's species of plants, animals, and microorganisms will be destroyed over the next 25 years because of rain forest destruction.
- One and one-half acres of rain forest are destroyed every second.
- Rain forests covered about 14 percent of the Earth's land surface in the past; today they cover only about 6 percent. Experts predict that in 40 years there may be no more rain forests left on Earth.
- Scientists estimate that 137 plant, animal, and insect species are lost every day due to deforestation, or 50,000 species a year. This represents a tremendous loss in potential life-saving medicines. Today 121 prescription drugs sold worldwide come from plant-derived sources from the rain forest. In the developed countries, one-quarter of the pharmaceuticals are derived from rain forest ingredients; but to date, less than 1 percent of the tropical trees and plants have even been studied to determine what their potential medicinal qualities are. Plants are becoming extinct faster than scientists can discover them.
- More than 20 percent of the Earth's oxygen is produced in the Amazon rain forest.

tion, condenses into clouds, and falls back to Earth again as rain in a repetitive cycle. The *evaporation* of the water from the Earth's surface also acts to cool the Earth. As climatologists continue to learn about the Earth's climate and the effects of global warming, they are able to build better models. When the tropical rain forests are replaced by agriculture or grazing, many climate models predict that these types of land use changes will perpetuate a hotter, drier climate. Models also

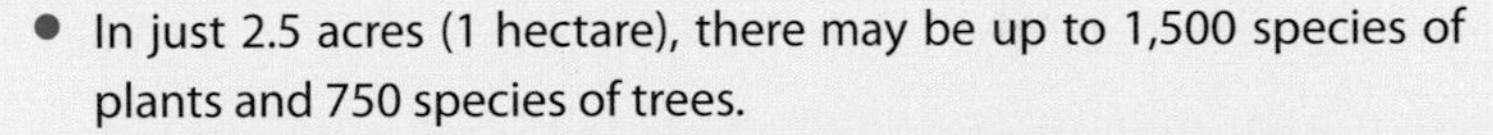

- In just 2.5 acres (1 hectare), there may be up to 1,500 species of plants and 750 species of trees.
- Approximately 80 percent of the developed world's diet originated in the tropical rain forest. The following foods originated from the rain forest:

avocados	figs	potatoes
black pepper	ginger	rice
cayenne	grapefruit	sugarcane
chocolate	guava	tomatoes
cinnamon	lemons	turmeric
cloves	mangos	vanilla
coconuts	nuts	yams
coffee	oranges	
corn	pineapples	

- The U.S. National Cancer Institute has identified 3,000 plants that are active against cancer cells. Nearly three-fourths come from the rain forest.
- In 1983, there were no U.S. pharmaceutical manufacturers involved in research programs to discover new drugs or cures from plants in the rain forest. Today there are more than 100 pharmaceutical companies involved in research along with several branches of the U.S. government, including Merck and the National Cancer Institute. Research projects include topics such as cancer, infections, viruses, Alzheimer's disease, and AIDS.

predict that tropical deforestation will disrupt the rainfall patterns outside the Tropics, causing a decrease in precipitation—even to far-reaching destinations such as China, northern Mexico, and the south-central United States.

Predictions involving deforestation can get complicated in models, however. For instance, if deforestation is done in a "patchwork" pattern, then local isolated areas may actually experience an increase in rainfall by creating "heat islands," which increase the rising and convection of air that causes clouds and rainstorms. If rainstorms are concentrated over cleared areas, the ground can be vulnerable and susceptible to erosion.

The carbon cycle plays an important role in the rain forests. According to NASA, in the Amazon alone, the trees contain more carbon than 10 years' worth of human produced greenhouse gases. When the forests are cleared and burned, the carbon is returned to the atmosphere, enhancing the greenhouse effect. If the land is utilized for grazing, it can also be a continual source of additional carbon. Deforestation changes the local weather. Cloudiness and rainfall can be greater over cleared land than over intact forest.

It is not certain today whether or not the tropical rain forests are a net source or sink of carbon. While the vegetation canopies hold enormous amounts of carbon, trees, plants, and microorganisms in the soil also respire and release CO_2 as they break down carbohydrates for energy. In the Amazon alone, enormous amounts of CO_2 escape from decaying organic matter in rivers and streams that flood huge areas. When tropical forests remain undisturbed, they remain essentially carbon neutral; but when deforestation occurs, it contributes significantly to the atmosphere.

Rain forest countries need to give serious consideration to future decisions. Currently, Papua New Guinea, Costa Rica, and several other forest-rich developing countries are seeking financing from the global carbon market (such as the United States) to create attractive financial incentives for tropical rain forest conservation. Developed nations helping developing nations is one approach to help getting a handle on the problem, but developed countries also need to cut back on their own emissions.

CLIMATE CHANGE IMPACTS

There are several impacts expected to occur in forested areas due to global warming, including increased timber production in some areas; regional variability; changes in nontimber production; impacts with fire, insects, and extreme events; changes in forest health; social and economic impacts; and changes in biodiversity.

According to the IPCC, current modeling efforts show that climate change will increase global timber production if there is an increase in forest growth as a result of higher CO_2 concentrations in the atmosphere, as well as a northward expansion of forest habitat in the Northern Hemisphere as temperatures warm and ecosystems migrate poleward. Economists in the United States do not predict a significant economic impact to forestry as a result of this poleward shift of forests due to the larger reserves of existing forests, technological change in the timber industry, and the ability to adapt in the United States.

Although the IPCC does predict an overall increase in timber production, they do caution that there will be regional variability due to local factors such as precipitation, temperature, humidity, soil type, and severe weather events. In California, for example, they project that initially (around 2020) climate change will increase harvests by causing an increased growth; but by 2100 there will be reductions in areas of softwood growth. They also predict there will be substantial negative impacts to other products from forests, such as seeds, nuts, resins, plants used in pharmaceutical and botanical medicine, cosmetics, and hunting resources.

The IPCC has also identified the need to further evaluate, revise, and fine-tune climate models and their predictions for forests, specifically the effect of CO_2 on tree growth. Some scientists think that projections of future tree growth may be overestimated in forest growth models. In one model devised by Richard J. Norby of the Environmental Sciences Division at the Oak Ridge National Laboratory, a 550 ppm CO_2 was used in the model, which yielded a net increase in productivity of 23 percent in young tree stands; but when this model was compared to an actual 100-year stand, there was not that much growth evident, suggesting the model greatly overestimated the accelerated growth effect of CO_2. Some scientists with the IPCC have suggested that other factors may also play roles that have not been accounted for in the model, such as land disturbances, air pollutants, and nutrient limitations.

Other identified limitations with current models include a need to include key ecological processes. What are needed are Dynamic Global Vegetation Models (DGVMs) to enable better predictions of climate-induced vegetative changes by simulating forest biomass, composition of species types, water and nutrient cycling, and the effects of wildfires.

Current models today indicate that increased temperatures and longer growing seasons will increase both aridity and additional fire risk. Models for the United States have predicted a 10 percent increase in the severity of fire. Other models, using a three-times greater increase in CO_2, project a 74 to 118 percent increase in area burned in Canada by the end of the 21st century.

In many forests, one of the most pressing issues is forest health. Even today, due to warmer temperatures, many forests in the western United States have experienced great infestations of bark beetle that have killed off thousands of acres of previously healthy forest. As global warming continues to intensify, pest and disease outbreaks will continue to be major sources of natural disturbance.

According to the IPCC, the effects will range from defoliation and growth loss to timber damage to massive forest diebacks. Climate change will shift current boundaries of insects and pathogens and also modify tree physiology and tree defense. This is one area that still needs improvement in modeling.

Other potential impacts to forests identified by the IPCC that will vary on a region-by-region basis include reduced access to forest land; increased costs for road and facility maintenance; damage to trees by wind, snow, frost, and ice; higher risks of wildfires and insect outbreaks; and the negative economic effects of wetter winters and earlier thaws to the logging industry.

Another way models need to be improved is by being able to portray the interactions among these multiple disturbances. As one impact occurs, it weakens the forest and increases its vulnerability so it is then even more susceptible to damage from other impacts. For instance, if strong winds damage trees by breaking branches, crowns, and trunks, it leaves the stand more vulnerable to damage from insect outbreaks and wildfires. In other cases, such as in the southwestern United States, prolonged drought increases species mortality, leaving these areas prone to insect and pathogen damage and wildfires. The Amazon is another

Wildfire is one of the projected negative impacts to forests as global warming continues. *(BLM)*

area where this is a major potential problem. Current deforestation, forest fragmentation, wildfire, and increased drought episodes make this region vulnerable to even more massive deforestation if the climate becomes warmer and drier.

According to data from the IPCC, social and economic impacts through the relocation of forest economic activity (as forests migrate) will affect businesses, landowners, workers, consumers, governments, and tourism. Areas where production increases will receive benefits; areas where there are decreases will have negative impacts. In areas where wood prices decrease, consumers will benefit while producers will suffer losses.

According to the FAO, people who will lose the most as far as forest resources are concerned are those who live in extreme poverty in developing countries and depend on forest resources for their livelihood, approximately 1,080,000,000 people. For these people, their income is tied directly to not only wood products but also to nontimber forest products as well, such as fuel, forest foods (nuts, fruits), and medicinal plants. Because forests also provide for the health needs of about 80 percent of the population, a health-related crisis threatens these communities. When a medicine man dies in the rain forest, it is like losing thousands of years worth of knowledge about the medicinal qualities of the plants in the rain forest.

Forest biodiversity will also be affected by global warming. Worldwide, there are potentially from 30 to 80 million species of plants and animals. The tropical rain forests, which cover only 7 percent of the Earth's landmass, provide habitat for more than half of them. Currently, only about 1.5 million species have been identified and named, leaving an enormous amount of life-forms still waiting to be discovered and studied. Many of these species are so specialized that they exist in very small, isolated ecosystems. This fact makes them extremely vulnerable to extinction in light of deforestation; if their habitat is cut down, they are eliminated. According to N. E. Stork of the Rainforest Action Network, up to 137 species are disappearing worldwide per day.

This represents not only an alarming loss of biodiversity, but also a potential source of medicine that could ultimately end up being a cure for life-threatening diseases such as cancer, AIDS, or many other serious diseases. Within the ecosystem itself, many organisms are losing species that they depend on, which means they also face extinction within the ecosystem, potentially causing a far-reaching ripple effect as *food chains* and biomes disintegrate.

AFTER DEFORESTATION AND THE FUTURE

After deforestation, the way the land is managed has important implications toward the land's future health—not only the soil and vegetation but also the life that inhabits the ecosystem. In a tropical rain forest, almost all of the nutrients are located in the plants and trees, not the soil. In the *temperate* and boreal forests, the opposite is true.

In tropical rain forests, when the trees are cut down in order to convert the land use to agriculture, farmers often burn the tree trunks to release the nutrients from the trees into the soil to grow the crops—a process called "slash and burn" agriculture. Over time, however, rain washes the nutrients from the soil, leaving the land infertile after a few years. At this point, the farmer finds a new plot of land to clear the forest from and starts over, abandoning the prior disturbed area. Unfortunately, rejuvenation of the forest is extremely slow because of the soil's lack of nutrients. It typically takes about 50 years for the forested area to grow back.

According to experts at Michigan State University in their rain forest research program, other types of farming are even worse for forest regrowth. Intensive agricultural systems such as plantations that use a lot of chemicals and fertilizers, kill animals, change the water balance, and weaken entire ecosystems, causing regrowth to take centuries.

Commercial logging is also a problem. Using a method called selective logging, where only selected trees are cut may seem nonintrusive, but because heavy machinery is used, many additional trees are damaged in the process. In a study conducted in Indonesia by Andrew Johns with Michigan State University and the NASA Earth Science Information Partners (ESIP), in an area where only 3 percent of the trees were logged, the logging operation actually damaged 49 percent of the trees in the forest. The positive side of selective logging, however, is that forest regeneration is much more rapid because there are still enough trees left to provide seeds and protect young trees from too much intense sunlight.

Logging using the clear-cutting method is much more damaging because all the trees are cut down, leaving the ground bare. Because even the tree trunks are removed, there are no nutrients left behind. These areas are so infertile, scientists do not know how long it will take for them to regenerate or if they even can. The following table illustrates the various types of tropical deforestation and the regrowth times associated with them.

A positive factor for temperate and boreal forests is that most of them are located in developed countries and are already well managed. One of the future impacts, however, that needs to be considered is the impact to, and adaptation of, wildlife to climate change. According to a study in *National Geographic News* on April 12, 2006, global warming could threaten up to one-fourth of the Earth's plant and vertebrate animal species with extinction by 2050. The study analyzed 25 biodiversity "hot spots" worldwide, such as the Floristic Region in the South African Cape, the Caribbean Basin, and the tropical regions of the Andes.

Methods of Tropical Deforestation

ACTIVITY	FACTORS	REGROWTH TIME
Slash and Burn Agriculture	Abandoned rapidly	Less than 50 years
Perennial Shade Agriculture	Some trees left	20 years
Intensive Agriculture (Banana Plantation)	Many pesticides, alteration of hydrology	More than 50 years
Cattle Pasture	Degradation of soils	More than 50 years
Selective Logging	Few trees cut	Less than 50 years
Clear-cut Logging	No trees or nutrients	More than 50 years

Source: ESIP

Wildlife habitat will be negatively affected as global warming continues, causing forests to migrate. If species cannot adapt to ecosystem changes, they face extinction. *(NPS)*

According to the study's lead scientist, Jay Malcolm of the University of Toronto, "These hot spots are the crown jewels of the planet's biodiversity. Unless we get our act together soon, we're looking at committing ourselves to this kind of thing."

The study was based on the assumption that CO_2 levels would double (as they are expected to do by 2100) and that temperatures would rise. The study projected that in the 25 hot spot regions, potentially 56,000 plant and 3,700 animal species could become extinct. Another independent study, conducted in 2004 by scientists from Conservation International, obtained similar results. They projected the possibility that more than a million species could be at risk of extinction by 2050.

According to Stuart Pimm, a Duke University expert in biodiversity and extinctions, "Species living in ecological hot spots are at particular risk when their environments change. That's where the most vulnerable species are, because they have the smallest geographical ranges. Species living high on tropical mountainsides, for example, have nowhere to go if temperatures warm their home turf. In South Africa's Cape Floristic

Region, located on the continent's southern tip, species are unable to migrate to lower latitudes to escape the rising temperatures."

Other experts warn that it is not just the "hot spots" that face imminent extinction risk. Terry Root of Stanford University's Center for Environmental Science and Policy says, "Many species are indeed struggling to hold on in locations all over the globe, not just in hot spots. This is not some activity that will be occurring "overseas." The likely extirpations and extinctions will also be occurring within a couple hundred miles of all of our back yards."

Based on a report in *LiveScience* on June 21, 2005, global warming is already changing animal habitats. According to Terry Root, "As humans argue about thermometer readings, animals are providing evidence that should be figured into scientific and political decisions. Animals are just reacting to what's going on out there. And if their behavior is very similar to what we expect with what's going on with global warming, we can use that information to support what the thermometers are showing."

The future will require solutions that will address multiple issues simultaneously—those involving timber and nontimber forest resources, wildlife, economic issues, human resources and living conditions, water quality, and other environmental concerns.

Anthropogenic Causes and Effects

While several natural processes add CO_2 to the atmosphere, the most significant source today in light of global warming is the human contribution—the anthropogenic effect. Until this source of CO_2 is brought under control and regulated worldwide, global warming will continue to escalate, worsening each decade. This chapter looks at that dilemma. First, it examines the carbon footprints left by people and nature, then the effects that land use has on the environment. Next, it looks at the role of public land and the incurred responsibilities of the U.S. government. It will also explore issues closer to home: Which cities are at risk and what is facing society today with irreplaceable cultural losses, as well as the impacts to people and society that global warming will have. Finally, it examines the economic pressures now being felt by insurance companies as a result of climate change.

CARBON FOOTPRINTS

When looking for causes of global warming, it is easy to point fingers and put the blame on industry, other nations, transportation, deforestation, and other sources and activities. But the truth of the matter is that every person on Earth plays a part and contributes to global warming. Even simple daily activities—such as using an electric appliance, heating or cooling a home, or taking a quick drive to the grocery store—contribute CO_2 to the atmosphere.

Scientists refer to this input as a "carbon footprint." A carbon footprint is simply a measure of how much CO_2 people produce just by going about their daily lives. For every activity that involves the combustion of fossil fuels (coal, oil, and gas), such as the generation of elec-

CALCULATING YOUR CARBON FOOTPRINT

Carbon calculators can be easily accessed on the Internet, are easy to use, and provide quick results. The following is an example of how they work:

1. Information: Your geographic location.
   Reason: Your carbon footprint is partially determined by where you live. Some areas rely on energy (such as electricity) generated mainly from coal (a dirtier fuel), others from cleaner energy sources such as wind energy.
2. Information: The size of your household.
   Reason: This helps differentiate you from other contributors.
3. Information: The amount of electricity used in the home.
   Reason: This calculates the amount of CO_2 generated by taking the total electricity used and dividing it by the price of power in the area. This number is then multiplied by the area's emissions factor, which relates to the type of energy the area uses. It also factors in the use of natural gas, heating oil, or propane, and whether a household participates in any renewable energy programs such as wind energy or solar power.
4. Information: CO_2 produced from transportation—the annual mileage and your car's make, model, and year. This is then multiplied

tricity, the manufacture of products, or any type of transportation, the user of the intermediate or end product is leaving a carbon footprint. Of all the CO_2 found in the atmosphere, 98 percent originates from the burning of fossil fuels.

Simply put, it is one measure of the impact people make individually on the Earth by the lifestyle choices they make. In order to combat global warming, every person on Earth can play an active role by consciously reducing the impact of their personal carbon footprint. The two most common ways of achieving this is by increasing their home's energy efficiency and driving less. A carbon footprint is calculated (carbon footprint calculators are available on the Internet), and a monthly, or annual, output of total CO_2 in tons is calculated based on the specific

by the emissions factor of gasoline or diesel fuel, which converts it to pounds of CO_2. For air travel, mileage is assessed along with takeoffs and lengths of flights.

Reason: This calculates the car's fuel efficiency.

Once all the figures are compiled, a total CO_2 output in tons is calculated—this is the carbon footprint. These can then be compared to national and global averages and used as a personal measuring stick to improve one's own efficiency in the use of carbon and contribution to global warming.

The following Web sites feature easy-to-use carbon footprint calculators:

www.carbonfootprint.com/calculator.aspx
www.climatecrisis.net/takeaction/carboncalculator/
coolclimate.Berkeley.edu/
www.3degreesinc.com/carbon_calculator/
www.terrapass.com/carbon-footprint-calculator/

(All have been accessed as of May 22, 2009.)

daily activities of that person. The goal then is to reduce or eliminate carbon footprints. Some people attempt to achieve "carbon neutrality," which means they cut their emissions as much as possible and offset the rest. Carbon offsets allow one to "pay" to reduce the global greenhouse total instead of making personal reductions. An offset is bought, for example, by funding projects that reduce emissions through restoring forests, updating power plants and factories, decreasing the energy consumption in buildings, or investing in more energy-efficient transportation. This can be an effective way to "plan" a carbon footprint.

In order to educate and make people more environmentally conscious, some companies now advertise what their carbon footprint is, drawing in additional business support because of their positive environmental commitments. Some commercial products now contain "carbon labels" estimating the carbon emissions that were involved in the creation of the product's production, packaging, transportation, and future disposal.

Carbon footprints are helpful because they allow individuals to become more environmentally aware of the implication of their own choices and actions and enables them to adopt behaviors that are more environmentally friendly—what is referred to as "going green." For example, transportation in the United States accounts for 33 percent of CO_2 emissions. Ways to make a difference include driving less, using public transportation, carpooling, driving a fuel-efficient car such as a hybrid, or bicycling. According to the Environmental Protection Agency (EPA), home energy use accounts for 21 percent of CO_2 emissions in the United States. Therefore, cutting down in these areas by increasing energy efficiency helps lessen the carbon footprint. There are several practical ways to do this: lowering the thermostat, installing double-paned windows, and installing good insulation, to name a few. Using compact fluorescent lamps and using energy-efficient appliances (such as those listed on the ENERGY STAR® program) also increases efficiency and lowers the carbon footprint. Carbon footprints can be a helpful measurement for those who want to take personal initiative and do their part to fight global warming.

This is the time for individuals to become aware of their personal behavior and how it affects the environment. The atmospheric concentration of CO_2 has risen by more than 30 percent in the last 250 years. Based on a study conducted by Michael Raupach of the Commonwealth Scientific and Industrial Research Organisation in Australia, worldwide carbon emissions of anthropogenic CO_2 are rising faster than previously predicted. From 1990 to 1999, the increase in CO_2 levels averaged about 1.1 percent per year; but from 2000 to 2004, levels increased to 3 percent per year. For this research, the world was divided into nine separate regions for analysis of specific aspects such as economic factors, population trends, and energy consumption. The result of the study showed that the developed countries (like the United States), which account for only 20 percent of the world's population, accounted for 59 percent of the anthropogenic global emissions in 2004. The developing nations were responsible for 41 percent of the total emissions in 2004, but contributed 73 percent of the emissions growth that year. The developing countries, such as India, are expected to become the major CO_2 contributors in the future. Today the largest CO_2 emitter is China.

This study is significant because even the IPCC's most extreme predictions underestimate the rapid increase in CO_2 levels seen since 2000. The scientists involved in the study believe this shows that no countries are decarbonizing their energy supply and that CO_2 emissions are accelerating worldwide.

Also from Australia's Commonwealth Scientific and Industrial Research Organisation, Josep G. Canadell calculated that CO_2 emissions were 35 percent higher in 2006 than in 1990, also a faster growth rate than expected. Canadell attributed this to an increased industrial use of fossil fuels and a decline in the amount of CO_2 being absorbed by the oceans or sequestered on land.

According to Canadell, "In addition to the growth of global population and wealth, we now know that significant contributions to the growth of atmospheric CO_2 arise from the slowdown of nature's ability to take the chemical out of the air. The changes characterize a carbon cycle that is generating stronger-than-expected and sooner-than-expected climate forcing."

In response to the study, Kevin Trenberth from the National Center for Atmospheric Research said, "The paper raises some very important issues that the public should be aware of: namely that concentrations of

CO_2 are increasing at much higher rates than previously expected and this is in spite of the Kyoto Protocol that is designed to hold them down in western countries."

Feedback on the study also came from Alan Robock from the Center for Environmental Prediction at Rutgers University in New Jersey. He said, "What is really shocking is the reduction of the oceanic CO_2 sink," referring to the ability of the world's oceans to absorb great amounts of CO_2 from the atmosphere. He also said, "It turns out that global warming critics were right when they said that global climate models did not do a good job at predicting climate change. But what has been wrong recently is that the climate is changing even faster than the models said. In fact, Arctic sea ice is melting much faster than any models predicted, and sea level is rising much faster than IPCC previously predicted."

According to the Brookings Institution, a research organization in Washington, D.C., America's carbon footprint is expanding. As America's population grows and cities expand, people are building more, driving more, and consuming more energy, which means they are emitting more CO_2 than ever before. The Brookings Institution believes that existing federal policies are currently limiting. They believe federal policy should play a more powerful role in helping metropolitan areas so that the country as a whole can collectively shrink its carbon footprint. They believe that besides economy-wide policies to motivate action, five targeted policies should be put in place that are extremely important within metro areas. They are identified as follows:

- promote a wider variety of transportation choices;
- design more energy-efficient freight operations;
- require home energy cost disclosure when selling a home in order to encourage more energy-efficient appliances in homes;
- use federal housing policies to create incentives to build with both energy and location conservation in mind;
- challenge/reward metropolitan areas to develop innovative solutions toward reducing carbon footprints.

The figure illustrates the results of the study conducted by the Brookings Institution. Of 100 metropolitan areas studied, the highest per cap-

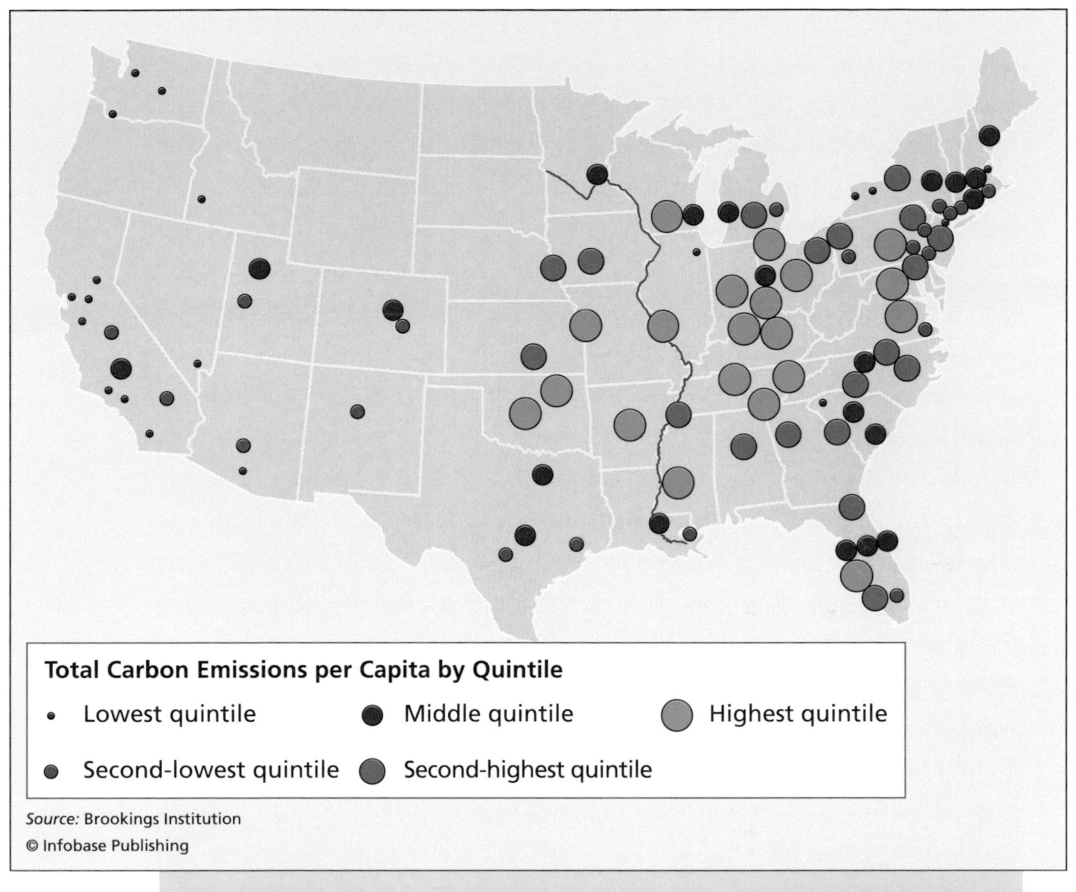

Per capita carbon emissions from transportation and residential energy use in the United States in 2005 *(Source: Brookings Institution, 2008)*

ita carbon emissions from transportation and residential energy use in 2005 was Lexington-Fayette, Kentucky, at 3.455 metric tons; the lowest was Honolulu, Hawaii, at 1.356 metric tons.

LAND-USE CHANGE

Land use and land cover affect the global climate system through biogeophysical, biogeochemical, and energy exchange processes. Variations in these processes due to land-use and land-cover change affect the climate. There are several natural processes that cause the exchange of greenhouse gasses such as CO_2 from the Earth's surface to the atmosphere, such as photosynthesis, respiration, and *evapotranspiration.*

Changing the land use of an area can also change the amount of heat transferred between the Earth's surface and the atmosphere. The nature of the Earth's surface is also important whether the surface reflects or absorbs heat. Reflectivity of the land's surface determines how much of the Sun's energy is absorbed and available as heat. Vegetation transpiration and surface hydrology determine how the heat will be spent—whether it is converted to latent heat (absorbed or radiated heat at constant temperature) or sensible heat (heat stored as a result of an increase in temperature).

There are also physical reactions transpiring at the Earth's surface. Both vegetation and land structure determine the surface roughness, which directly determines air momentum and heat transport. According to the U.S. Global Change Research Program (USGCRP) land-use and land-cover change studies are also important because large-scale vegetation biomass and forest cover provide a critical component to the carbon cycle. Understanding how land-use change affects this biomass and how the subsequent changes in the carbon cycle affect global warming—the linkages and feedbacks between land-use and land-cover change, climate change forcings, and other related human and environmental components—is critical in order to slow global warming today and in the future.

The USGCRP also believes: "Research that examines historic, current, and future land-use and land-cover change, their drivers, feedbacks to climate, and environmental, social, economic, and human health consequences is therefore of utmost importance and often requires interagency and intergovernmental cooperation."

One of USGCRP's examples of a multiagency effort is their Congo Basin Forest Partnership, which focuses on conserving the second-largest tropical rain forest in the world in equatorial Africa. They are currently using satellite data to map forest extent, determine habitat fragmentation, and enforce conservation laws in order to minimize greenhouse gas emissions as a result of deforestation.

Another project they were directly involved in was the North America Land Cover Summit that was held in September 2006, which was designed to gather information and encourage collaboration among private institutions and government agencies in order to advance the development and application of land-cover information in Mexico, Canada, and the United States.

In 2008, they were involved in research dealing with potential land-use and land-cover change patterns, dynamics, and drivers; understanding the mutual effects and feedback between climate variability and land use/land cover; and forecasting environmental, social, economic, and human health consequences. They used satellite imagery to accomplish this.

Even though the buildup of CO_2 in the atmosphere is one of the best-known anthropogenic contributions to global warming, variations in land use and surface cover are also significant impacts that affect climate change. According to the Union of Concerned Scientists (UCS), activities such as urban sprawl, agriculture, deforestation, and other human influences have altered and fragmented the natural landscape enough that it has changed the global distribution of CO_2 and has changed the Earth's energy balance, which can affect the climate on multiple scales. Scientists at UCS have determined that even relatively small changes in land use affecting only 100 square miles of urban development or deforestation can change the local rainfall pattern enough that it can trigger climate disruptions.

According to a study conducted by Eugenia Kalnay and Ming Cai of the University of Maryland, land-use changes are responsible for more of the rise in global temperatures than scientists previously had thought.

According to Kalnay, who is a professor of meteorology, "Our estimates are that land-use changes in the United States since the 1960s resulted in a rise of more than 0.2°F (0.1°C) in the mean surface temperature, and estimate twice as high as those of previous studies. These findings for the United States suggest that land-use changes may account for between one-third and one-half of the observed surface global warming. The larger effect found in this study is likely because our method covers all changes in land use. Previous methods for estimating the impact of land-use change relied on measures of light at night—that only provide an indication of the effects of urbanization, but not of other changes in land use."

According to research conducted at the University of Guelph, land-use modifications for urbanization and agriculture have affected climate change data more than was previously thought. When land is changed from forests to seasonal crops or from natural land cover to urban environments, regional climate systems can be altered. Clear-cut hillsides, for instance, are much warmer than natural forested hillsides. In areas of heavy urbanization they become "heat islands" because so much heat energy is generated from industry, transportation, cars, and asphalt's absorption of solar energy. Deforestation, urbanization, and agricultural practices also add CO_2 to the atmosphere.

The National Center for Atmospheric Research (NCAR) in Boulder, Colorado, has completed simulations of 21st-century climate that show that human-produced changes in land cover could produce additional warming in the Amazon region comparable to that caused by greenhouse gases. The results also showed that changes in land use in some midlatitude areas could counteract some greenhouse warming. NCAR has concluded from their simulations that including land cover as a variable in their computer modeling of climate change is important in order to get a true reflection of the future.

Johannes Feddema, of the University of Kansas, who led the research included for the first time variables representing not only the atmospheric and oceanic components of global warming but also changes in land cover as a result of deforestation, urbanization, agriculture, and other human activities.

In their simulations, deforestation added 3.6°F (2°C) or more to surface temperatures in the Amazon by 2100. In addition, it also showed a cooling in the nearby Pacific and Atlantic waters with a weakening of the large-scale Hadley circulation. Converting midlatitude forests and grasses to croplands had a cooling effect because crops reflect more sunlight and release more moisture to the air. The research team found that converting more land to agriculture tends to counteract global warming up to 50 percent across parts of North America, Europe, and Asia. The boreal forests in Canada and Russia, however, add to regional warming as they migrate northward.

According to Feddema, "The choices humans make about future land use could have a significant impact on regional and seasonal climates. The purpose of our project is to include human processes more directly in global climate models. This is the first step."

PUBLIC LANDS

In the United States, one-third of the land is owned by the public and administered by the federal government—namely, the National Park Service, the U.S. Forest Service, and the Bureau of Land Management. The total area size of public land is roughly 700 million acres (283 million hectares). These areas are protected and managed for different purposes—some, such as designated wilderness areas, are kept pristine and natural because of their inherent beauty, exceptional quality, or unique habitat. Others, such as national parks, are managed in ways to preserve their unique character and also allow tourists (the public) to enjoy them. Other lands are managed for a multitude of different uses for public entertainment and recreation, such as campgrounds that support fishing, hiking, boating, mountain biking, off highway vehicle (OHV) riding, and horseback riding.

These designated areas, however, are susceptible to events influenced by climate variability, such as drought, wildfires, severe storms, and poor air quality. According to the IPCC, global warming may cause an increase in the incidences of drought and wildfire. They also point out that as sea levels rise, public lands in coastal areas will be compromised and eroded and wetlands could become inundated, destroying natural habitat.

The National Park Service has already confirmed that some national parks are already feeling the negative effects of global warming. In Glacier National Park in Montana, many of the glaciers in the park that draw visitors from around the world are currently melting at an accelerated rate. In 1850, the park had 150 glaciers, but because of global warming, today there are only 27 glaciers left. These surviving glaciers are also rapidly shrinking—the largest of the remaining glaciers are only 28 percent of their previous size. Alaska is also experiencing the loss of several glaciers.

Protected areas in Florida, which draw millions of tourists for recreational activities such as scuba diving, are experiencing massive coral bleaching because of rising temperatures of ocean water, which kills the sensitive coral ecosystems. *El Niño* events have been blamed as the cause of some of these bleaching events.

In 2001, the U.S. Fish and Wildlife Service (FWS), the National Park Service (NPS), and the U.S. Environmental Protection Agency (EPA),

The Muir Glacier located in Alaska is an example of the glacier melting that is occurring worldwide as a result of warmer temperatures—the top image was taken in 1941, the bottom in 2004. *(William O. Field [top], Bruce F. Molnia [bottom], National Snow and Ice Data Center)*

formed a multiagency partnership to create the Climate Change, Wildlife and Wetlands Toolkit for Teachers and Interpreters in a national effort to educate the general public about global warming and the negative effects it will have on the landscape and ecosystems in parks, wildlife refuges, wilderness areas, and other areas reserved and protected for restricted, specific, endemic, and endangered species in the event of climate change if biomes shift poleward with warming temperatures.

In order to further this effort, in 2003, the NPS and EPA created an additional program called Climate Friendly Parks (CFP). Through the efforts of this program, the participating agencies are finding ways to reduce emissions from park activities and also educate the public about potential impacts in the park and what proactive action the government is taking to solve any environmental problems encountered due to the operation of the park. For example, in the past at Zion National Park in Utah, private cars were allowed to travel the routes inside the park to the different geological attractions. It got to the point, however, that thousands of cars per day were navigating the narrow roads between geologic areas in the Zion Canyon Scenic Drive, causing not only heavy congestion but also pollution. The NPS then decided to restrict private travel within that extremely popular portion of the park and limit visitation to touring shuttles only.

The nation's national parks offer multiple resources, outdoor recreational activities, breathtaking scenery, and many tourism opportunities. These activities, however, are extremely sensitive to changes in temperature, rainfall, snowstorms, drought, and wildfire. In fact, they are sensitive to any climate variability and change. Within these protected areas, species of wildlife and vegetation have adapted to specific climatic conditions that allowed them to thrive in their respective environment. Some life-form biomes are extremely small and specialized. For instance, a single tree in a rain forest biome can harbor 30,000 different species. Because of this, these specialized biomes face an extremely difficult challenge if their biome is presented with changing conditions due to global warming. In many cases, if the change is too great, the species will not survive because they and/or the biome will be unable to navigate and relocate in time.

Grand Teton National Park provides a home to a great diversity of wildlife and draws millions of visitors. *(NPS)*

According to the IPCC, the effects of climate change on tourism in a particular area depend in part on whether the tourist activity is summer- or winter-oriented and, for the latter, the elevation of the area and the impact of climate on alternative activities. Therefore, the range of effects on each area will differ depending on the global warming effects at that particular location. As an example, an area today that serves as a ski resort may not have snow in the future, so may have to become a mountain resort used for other activities such as hiking, mountain biking, or horseback riding.

According to the Outdoor Industry Foundation in their August 2006 report, the outdoor recreation industry has a $730 billion impact on the U.S. national economy. They reported, "This amount factors in the amount Americans spend on outdoor trips and gear, the companies that provide that gear and related services, and the companies that support them. The outdoor industry also supports 6.5 million jobs, which amounts to one in 20 U.S. jobs, and generates about $88 billion in federal and state tax revenue while stimulating 8 percent of all consumer spending.

If global warming negatively affects the recreational uses of mountain areas, these changes will have a ripple effect of economic consequences throughout the local, national, and international economy, particularly in business sectors such as the travel industry, the transportation industry, sports equipment industry, food industry, and others related to leisure activities.

Some of the possible effects that will occur in the U.S. on recreation and tourism include:

- Declines in cold-water and cool-water fish habitat may affect recreational fishing.
- Shifts in migratory bird populations may affect recreational opportunities for bird-watchers and wildlife enthusiasts.
- Coastal regions may lose pristine beaches due to sea-level rise.
- Winter recreation (skiing, snowmobiling, ice fishing) is likely to be affected by reduced snowpack and fewer cold days, and costs to maintain these industries will rise.
- Tourism and recreation in Alaska will most likely undergo climate-driven transformation through loss of wetlands and a reduction in habitat for migratory birds.
- Arctic breeding and nesting areas for migration birds will be lost, affecting bird-watching activities.
- Melting permafrost in the Arctic will pose economic damage to structures and other infrastructures to residents there.

Wildfires are another impact as a result of global warming, some of it anthropogenically caused. In a study that appeared in *Science,* as reported in *USA Today* on July 7, 2006, not only the number, but the size of large forest fires in the West has grown "suddenly and dramatically" during the past 20 years because of global warming. Since 1987, the wildfire season has become extended 2.5 months longer than it traditionally had because springtime was seeing higher temperatures and winter snowpack was melting faster. With earlier snowmelt, the soil and vegetation dries out sooner, which leads to more fires that burn bigger and longer.

In a study conducted by the University of California-San Diego and the University of Arizona, 1,166 large forest fires were analyzed from 1970 to 2004 on national forest and park land in the western United States. From 1987 to 2004, there were four times as many forest fires and 6.5 times as much land burned as there was from 1970 to 1987.

According to Tom Swetham of Arizona's Laboratory of Tree-Ring Research, "Climate is the principal reason. The rising temperatures and dryness due to warming are the main factors in the northern Rockies."

Each year large amounts of federal funding are allocated toward fighting fires on public lands. As global warming continues, this effort is expected to intensify.

CITIES AT RISK

Cities are at risk as global warming escalates for two major reasons: sea-level rise and the urban heat island effect. Both are serious environmental challenges. According to the Environmental Protection Agency (EPA), global warming can have a profound effect on many components in cities: It can affect tunnels, communications lines, power plants, buildings, sewers, water systems, and every possible physical structure in the region. Because most Americans now live in urban and suburban areas (and by 2050, most of the world will), it is critical that cities find out what impacts could potentially occur and make plans now to deal with any future disasters in order to minimize the risks and possible damage.

Paul Kirshen, a research associate professor of civil and environmental engineering at Tufts University who has been involved with the EPA on a study focusing on the effects of global warming on cities, comments on the social impacts to society: "People may react differently in a warmer city, exacerbating some of those physical impacts. For instance, the warmer the weather, the more water and energy people use. So the infrastructure may get it in two different ways—the physical impacts, and the social and economic impacts."

Mindy Lubber of the EPA in the New England region says, "If we don't address these issues, the price we pay may be huge. Disruptions to infrastructure can be costly, as we in the Boston area learned during the flood of 1996. Extreme weather caused $70 million in property damage and disrupted transportation for thousands of people. The costs of modifying and repairing our infrastructure can be exorbitant."

Sea Level Rise

Global warming causes the oceans to warm and expand, causing sea levels to rise. If ocean levels rise high enough, shoreline areas will be inundated and lost, forcing the populations that inhabit these areas to relocate and causing billions of dollars of economic losses in property damage.

Coastal lowlands worldwide are some of the most rapidly growing population areas. The land area referred to as the low elevation coastal zone (LECZ) is the contiguous area along the coast that is less than 33 feet (10 m) above sea level. This zone covers only 2 percent of the Earth's total land area but provides a home for 10 percent of the world's population and 13 percent of the world's urban population—more than 600 million people live in this zone, and 360 million are urban. Included in this group are the world's small island countries and those large countries that have heavily populated areas within delta regions.

In a study conducted by the International Institute for Environment and Development in 2007, mitigation measures may be the best way to avoid the risks brought on by climate change (such as sea-level rise), but it is too late to rely solely on mitigation measures. They believe that migration of the population from these low-lying coastal areas is important, but it can be very expensive and cause major disruptions to peoples' lives. Adjustments that can be made to protect residents, however, should be accomplished as much as possible. It should be clearly understood, however, that low-income groups living on floodplains are especially vulnerable, and because any mitigation to solve the problem is a lengthy process and takes a long time to complete, any future urban development in more suitable areas should be considered instead.

Settlement along coasts is not always easy to control, however. Historically, coasts were desirable places to settle because of their immediate access to trade and business. Today, as in the past, the recent expansion of international trade has also encouraged people to move to coastal locations. As an illustration, today China has one of the largest trade-related coastward movements the business world has ever seen.

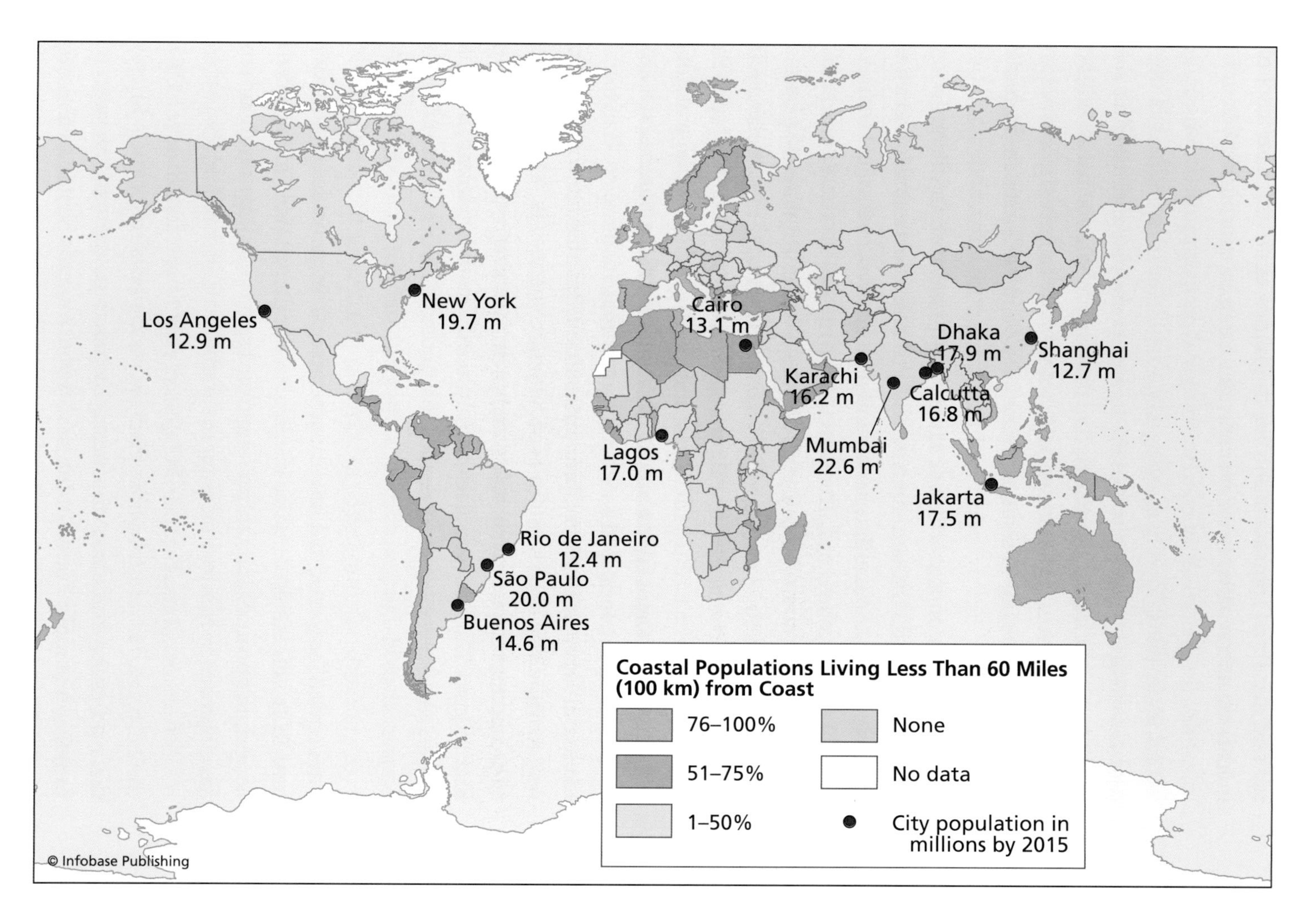
Los Angeles
12.9 m
New York
19.7 m
Cairo
13.1 m
Karachi
16.2 m
Mumbai
22.6 m
Dhaka
17.9 m
Calcutta
16.8 m
Shanghai
12.7 m
Jakarta
17.5 m
Lagos
17.0 m
Rio de Janeiro
12.4 m
São Paulo
20.0 m
Buenos Aires
14.6 m
Coastal Populations Living Less Than 60 Miles (100 km) from Coast
76–100%
51–75%
1–50%
None
No data
City population in millions by 2015
© Infobase Publishing

If global warming causes sea levels to rise, millions of people will be negatively affected.

With global warming, the world's coastal populations are at risk from flooding, especially when high tides combine with storm surges and/or high river flows. According to the International Institute for Environment and Development, "Between 1994 and 2004, about one-third of the 1,562 flood disasters, half of the 120,000 people killed, and 98 percent of the two million people affected by flood disasters were in Asia, where there are large population agglomerations in the flood-plains of major rivers (such as the Ganges-Brahmaputra, Mekong, and Yangtze), and in cyclone-prone coastal regions (such as the Bay of Bengal, South China Sea, Japan, and Philippines)."

Worldwide, developers have drained wetlands in order to convert those ecosystems into urban areas. The problem with doing this, however, is that it removes a "natural buffer" against tidal floods. In delta regions, land compaction and subsidence from groundwater withdrawal can lead to sea-level rise, which also increases flood risk. Unfortunately, it is the low-income areas that are the most vulnerable.

According to the IPCC, faster sea-level rise of more than 3.3 feet (1 m) per century could result from accelerated melting of the Greenland ice sheet or the collapse of the West Antarctic ice sheet (although not likely during the 21st century). The IPCC estimates that without any other changes a sea-level rise of 15 inches (38 cm) would effect five times more people flooded by storm surges.

From a management perspective, the International Institute for Environment and Development believes that for these areas to fare better with global warming, they need to have effective coastal zone

(opposite page) Coastal erosion, saltwater intrusion into freshwater supplies, and coastal storms are all events that can harm developed coastal areas because of global warming. Because coastal areas are attractive areas to live, millions of people may face physical and economic disasters as global warming continues.

management applied to them. Because structures are permanent and cannot be relocated, taking potential risks into account before heavy development begins in new areas is important. Sometimes relatively small shifts in settlement location, such as out of a coastal plain onto more elevated ground, can make a major difference later on.

One area that is especially vulnerable is Bangladesh. According to Dr. Robert Buddemieir of the Kansas Geological Survey, "Bangladesh is massively populated, achingly poor, and something like a sixth of the country is going to go away." Bangladesh cannot afford to build barriers to hold back the sea, so people would have to move inland, increasing the population's density and leading to an increase in hunger and disease.

The Maldives in the Indian Ocean have the same problem. Consisting of 1,190 separate islands with an average elevation of five feet (1.5 m) above sea level, if the sea rises, more than 200,000 people would be forced to abandon their homes.

Urban Heat Island Effect

The "urban heat island effect" is a phenomenon created by urban surfaces such as pavements (like asphalt) and conventional roofs that absorb the Sun's energy and reradiate it as heat. This can increase the temperature of a city several degrees higher than that of surrounding rural areas. The additional heat generated by running air conditioners and machinery, and from vehicle exhaust further worsens the urban heat island effect.

In a NASA study conducted by Stuart Gaffin, an associate research scientist with the Earth Institute at Columbia University, he outlined two important reasons to be concerned about urban heat islands. The first reason is the established trend of the increase in numbers of city dwellers.

According to Gaffin, "About half the world's population—three billion people—now live in cities. In a couple decades, it's going to be five billion people."

He says that in developing countries people often migrate to cities not for economic reasons such as a lucrative job market or business opportunities, but because of crop failures, natural disasters (such as flooding or drought), or to escape warfare. Because of this, city dwellers

AREAS MOST AT RISK

About 634 million people—10 percent of the global population—live in coastal areas that are within just 33 feet (10 m) above sea level, where they are most vulnerable to sea-level rise and severe storms associated with climate change. Three-quarters of these are in the low elevation coastal zones in Asian nations on densely populated river deltas, such as India. The others most at risk are the small island nations. According to the Earth Institute at Columbia University, the 10 countries with the largest number of people living within 33 feet (10 m) of the average sea level in descending order are as follows:

China	Indonesia	Thailand
India	Japan	Philippines
Bangladesh	Egypt	
Vietnam	United States	

The 10 countries with the largest share of their population living within 33 feet (10 m) of the average sea level are as follows:

Bahamas (88%)	Guyana (55%)	Egypt (38%)
Suriname (76%)	Bangladesh (46%)	Gambia (38%)
Netherlands (74%)	Djibouti (41%)	
Vietnam (55%)	Belize (40%)	

Over the last century, global sea level has risen on average of 4–10 inches (10–25 cm). Some estimates predict the rate of sea-level rise could double by the end of this century. If global warming continues unchecked, major urban areas built near sea level will see significant impacts. As an illustration, the area along the Eastern Seaboard of the United States, including New York City; Boston; Washington, D.C.; and Miami, will be at risk with an expected sea-level rise of 18–20 inches (46–51 cm) above current levels by 2100. Low-lying infrastructure in these areas, including buildings, roads, power lines, airports, train and subway systems, are all at risk of increased flooding. Popular recreational areas, such as the beaches of New Jersey, Long Island, North Carolina, and Florida's Gold Coast, will be at risk of accelerated beach erosion and loss. Wetland areas would also be threatened, endangering important habitat for shorebirds, plants, and nursery areas for fish, as well as valuable ecosystem services. Saltwater intrusion into underground water resources is also a serious issue, threatening water quality for residential and industrial users.

are often very poor and cannot afford luxuries like air-conditioning, refrigeration, or medical care. As a result, when heat waves strike—repeated days of extremely high temperatures—populations are dangerously vulnerable to health hazards.

The second reason for Gaffin's concern is climate change. Acknowledging global warming and the effects it is currently having on the environment, he believes that even if countries begin reducing their greenhouse gas emissions right now, a certain amount of further warming is inevitable because of the CO_2 and other greenhouse gases that have already been released into the atmosphere. He expects a warmer climate to make future heat waves more severe. He says, "I think mitigating global warming is important, but even with mitigation, all these people living in cities are going to experience some warming. Right now, we average about 14 days each summer above 90°F (54°C) in New York. In a couple decades, we could be experiencing 30 days or more. So we have two forces—urban heat islands and global warming—that are reinforcing each other and are going to create hot, hot conditions for more than half the world's population. How do we make cities more habitable in the future?" Gaffin bases his beliefs on research conducted by the U.S. Global Change Research Program.

One of the problems with heat retention in cities, besides dark surfaces absorbing heat from car motors, hot air from air conditioners and clothes dryers, machinery, and industrial smokestacks, is that they lack enough vegetation to offset the high temperatures. Vegetation cools its surroundings by evaporating water. According to the Environmental Protection Agency (EPA), it is common for urban areas to be 10°F (6°C) hotter than nearby suburban and rural areas.

In a study Gaffin conducted in New York City—which has created an urban heat island for more than a century—he accessed satellite temperature data, city-wide land cover maps, weather data, and a regional climate model to devise a plan to cool the city. The purpose of the study was to determine how much the city could be cooled by planting trees, replacing dark surfaces with lighter ones, and installing vegetation-covered "green roofs."

Using a model called "MM5" developed by Pennsylvania State University and the National Center for Atmospheric Research, they were able to predict regional atmospheric circulation and weather phenomena and determine that a combination of urban forestry and light roofs could reduce New York City's overall temperature by 1.2°F (0.67°C) throughout the day. At 3:00 P.M., when temperatures are usually at their highest, the temperature could be reduced 1.6°F (0.89°C). Power demands increase sharply as temperatures rise above 60°F (36°C).

Although these temperatures may not seem significant, they can have a significant impact. According to Gaffin, "Power usage is very sensitive to even slight variations in temperature. If you're near the maximum power availability of your grid, which is often the case in these summer heat waves, each degree can make a significant difference in overloading the grid and leading to a blackout. So anything that would cool the city off, even a little, would ease energy demands and reduce blackout risks."

Gaffin found the model helpful, but he still had some practical reservations about it: "A lot of regional climate models were originally designed for natural land surfaces. For cities we've got to take into account the hundreds of thousands of building geometries."

He does believe that planting trees is critical. Not only does each tree cool its immediate area, but it helps shade adjacent buildings. The study estimated that 17 percent of the city's surface could be planted with trees to help cool the environment.

The last major areas of consideration in the city were the impervious surfaces such as roads and adjacent areas that cover about 64 percent of the city.

"The tar beach roof is an oven in the summer, reached 160°F (96°C)," Gaffin explained. "So cooling of this roof is like turning off an open oven."

For roofs, there are two options: a light-reflecting surface or vegetation. Of the two, he decided a vegetated roof was the better choice for a couple of reasons. First, the reflection of light off the light surface may be redirected onto another building with a darker surface and be absorbed and converted to heat. Second, the light-colored surface would still be impervious to rainwater, causing it to run off that surface and enter sewer systems, overloading them and putting raw sewage into waterways and causing health problems.

Instead, Gaffin supports the concept of the vegetation-covered or green-roof system. In the study, a roof planted with *Sedum spurium* was compared to a standard dark roof.

"*Sedum* is a desert-adapted plant with shallow root systems. The plants can tolerate long periods of drought. They're lush green, beautiful to look at, and quite pleasant to touch," says Gaffin.

He and his team found that the peak temperatures on the roofs that were planted with *Sedum* were 54°F (30°C) lower than the temperatures on standard roofs. They also found that the drought-resistant vegetation was able to survive without watering.

Vegetation-covered roofs can also last many decades and actually leak less than standard roofs. They absorb water like a sponge, avoid runoff problems that are typical with standard roofs, and they evaporate water, cooling the air around them. They are beginning to grow in popularity in the United States, although cost is still somewhat prohibitive; but they have been used for centuries in Europe. They are also growing in popularity in Canada and Japan. In Tokyo, the municipal government requires all structures with roof areas greater than or equal to 10,765 square feet (1,000 m^2) to cultivate at least 20 percent of that roof area.

As temperatures continue to rise with global warming, however, the green roof's potential to cool cities in the United States is becoming more attractive, especially as death tolls add up from recent heat waves. The Chicago heat wave in 1995 killed more than 700 people, and the European heat wave of 2003 killed nearly 45,000 people.

Currently, planted roofs are now realizing positive benefits. In Chicago, the City Hall building is vegetated. On hot days, temperatures on that building are 25–80°F (14–44°C) cooler than the adjacent building.

According to Brad Bass with Environment Canada, "It's hard to estimate the exact amount, but green roofs can lead to at least a five percent, if not even a 15 percent, reduction in electricity in the summer."

The results of a study conducted by Environment Canada showed that if 6 percent of Toronto's roof area was converted to planted roofs, greenhouse gas emissions in the city would be reduced by 2.4 megatons a year. Most of the rain is absorbed by the plants and soil to later evaporate or transpire back to the atmosphere as water vapor, providing a cooling effect.

CLIMATE SOLUTIONS FOR CITIES

According to the Environmental Protection Agency (EPA), there are several municipal-level actions that can save money, save energy, clean the air, reduce congestion, curb sprawl, and reduce greenhouse gas emissions. They have identified the following:

- Make building energy improvements.
- Replace motors in city operations with more efficient models.
- Buy ENERGY STAR® equipment for municipal offices.
- Change traffic lights to light-emitting diode (LED) fixtures.
- Use renewable energy systems to improve air quality.
- Purchase green power to improve air quality.
- Redesign communities to encourage walking, biking, and mass transit.
- Provide incentives for mass transit or carpooling.
- Foster telecommuting and similar trip reduction programs.
- Convert fleets to run on alternative fuels.
- Put police on bicycles.
- Initiate "Pay As You Throw" waste disposal programs
- Implement curbside recycling.
- Recycle office paper and reduce landfill costs.
- Buy products made from recycled materials.
- Establish composting programs.
- Capture methane from landfills.
- Integrate Smart Growth in planning.
- Plant trees to keep buildings and streets cooler to improve air quality, lower air-conditioning loads, and save money.
- Use highly reflective surfacing and roofing materials.

Different approaches are used in different cities, usually determined by individual city government policies and budgets. Some are also easier to implement sooner than others. For example, when new equipment is needed, the energy-efficient models can be immediately purchased and used. Older, inefficient models, however, may continue to be used for several years until they either need to be replaced or an agency can afford to replace them. Planting trees, recycling materials, and purchasing

(continues)

(continued)

recycled products are relatively easy solutions that can be done immediately without much planning. Redesigning communities and converting fleets to use alternative fuels are more difficult tasks and take more effort to accomplish, often requiring government policies and budgets to be adjusted. With careful planning, however, all of these actions can be accomplished in order to reduce the impacts of global warming.
Source: EPA

CULTURAL LOSSES

Perhaps one of the saddest losses to the effects of global warming is the abuse and damage being suffered by the world's most prized monuments and works of priceless art—works often representing the unique cultural story of a people—irreplaceable works of heritage and history. Many of these sites are located in coastal areas, threatened by sea-level rise, flooding, and intense storms, posing threats to monuments, archaeological sites, and other famous structures. Because of climate change, other natural areas are losing the physical character that makes them noteworthy.

In Boston, for example, with continued global warming, sea-level rise and coastal storms could increase the height of flooding on the Charles River. If this were to happen, it would destroy famous historical sites such as Boston Public Garden. In Scotland, archaeologists estimate there are currently 12,000 sites whose existence is threatened by coastal erosion. If global warming causes sea levels to rise in this area, those sites may never be able to be explored and inventoried, essentially causing the loss of important elements of history.

In Venice, Italy, because of the combination of rising sea levels and ground subsidence, the canal system that dissects the city frequently floods many buildings. This is expected to worsen as global warming increases. Today St. Mark's Square floods about 50 more times each year than it did in the early 1900s. At the West Coast National Park in South Africa, many of the archaeological sites are being ruined due to sea-level rise, destroying the future discovery of ancient human history dating back 117,000 years. In the Czech Republic in 2002, flooding destroyed numerous concert halls, museums, libraries, and theaters. Climate change is expected to exacerbate this problem.

The Belize Barrier Reef, once described by Charles Darwin as "the most remarkable reef in the West Indies," is suffering from the effects of global warming. Classified as a World Heritage Site, the reef is being bleached because of rising water temperatures. Reefs worldwide are facing this fate, and as global warming continues, reefs will continue to perish.

According to a report on CBS News in London in June 2007, higher temperatures and humidity will speed up the corrosion of the Eiffel Tower's ironwork. Moist salty air will attack the fine stonework on buildings from the Parthenon to the Tower of London and Westminster Abbey. The end result will be that in a hundred years from now the fine detailing on the buildings will be lost. Because of the effects of rising temperatures, precious treasures that have lasted for centuries may be ruined in just a few decades from now.

ECONOMIC IMPACTS

According to a report in *LiveScience* in November 2005, insurance companies are starting to take notice of the destructive effects of global warming. As property damage occurs from flooding, rising sea levels, hurricanes, severe weather, and other global warming–related events, insurance companies will be increasingly called upon to compensate property owners for destroyed property. Health insurance companies will be faced with the same issues as people are injured in violent weather events or suffer from heat waves, allergies, infectious diseases, and starvation. As more people are affected, insurance companies will be harder hit to bear the financial burden of global warming.

One insurance company, Swiss Re, which had a study conducted by the Center for Health and the Global Environment at Harvard Medical School and also sponsored by the United Nations Development

Program, issued the comment: "Climate change will significantly affect the health of humans and ecosystems and these impacts will have economic consequences."

Through the completion of 10 case studies, they found that climate change will have multiple effects from challenges such as infectious diseases (malaria, West Nile virus) to severe and extreme weather events such as hurricanes, heat waves, and floods. Changes to a multitude of different land-use types were analyzed, such as forests, deserts, agriculture, range, and marine habitat. A range of cause-and-effect issues were considered that had direct implications for insurance companies.

According to Dr. Paul Epstein, the associate director of the Center for Health and the Global Environment at Harvard Medical School, "We found that impacts of climate change are likely to lead to ramifications that overlap in several areas including our health, our economy, and the natural systems on which we depend. Analysis of the potential ripple effects stemming from an unstable climate shows the need for more sustainable practices to safeguard and insure a healthy future."

In the study, economic implications and future impacts were analyzed in 10 different case studies. Issues were taken into consideration, such as ragweed pollen growth being accelerated by increasing levels of CO_2 in the atmosphere, which may be contributing to the documented rising in asthma cases. Swiss Re, the insurance company instigating the study, has been concerned about the insurance costs of global warming since 2003 and is concerned about the economic implications in the future.

Jacques Dubois of Swiss Re says, "Whereas most discussions on climate change impacts home in on the natural sciences, with little to no mention of potential economic consequences, this report provides a crucial look at physical and economic aspects of climate change. It also assesses current risks and potential business opportunities that can help minimize future risks."

As portrayed in this chapter, the human element is playing a significant role in global warming, and its contribution has the potential to affect millions of people worldwide. Unless it is controlled and slowed considerably, the physical, economic, and health effects have the potential to be globally devastating.

6

The Fate of Natural Refuges

Based on research and conclusions from the Intergovernmental Panel on Climate Change (IPCC), global warming is the greatest threat to wildlife today. This chapter examines the impact global warming is having now and in the future on nature and its wildlife if serious action is not undertaken to stop it now. The chapter will look at specific animals and the fate of wildlife refuges that are at risk because of climate change. It will also present legislation designed to protect wildlife against this very serious challenge.

CLIMATE CHANGE IMPACTS TO NATURE AND WILDLIFE

In a study published by *Nature,* human-induced climate change could result in the extinction of more than a million terrestrial species by 2050. The study looked at six biodiversity-rich areas around the world and estimated that 15 to 37 percent of the terrestrial species would be gone, depending on the rate of global warming.

Dr. Terry L. Root of the Center for Environmental Science and Policy at Stanford University in California, an organization involved in international environmental assessment, negotiation, remediation, and protection, conducted a study that provided concrete evidence that global warming is affecting the world's plants and animals by forcing them to shift their ranges. It is also affecting the timing of spring-related events such as egg laying and migration.

The Pied flycatcher, an avian species found in Europe, for example, is not arriving early enough in the spring because it is cued by length of day when to migrate. The problem is that spring is coming earlier to its migration destination due to warming temperatures. The earlier warming is causing its major food source, caterpillars, to become abundant earlier because their life cycle is governed by temperature, instead of length of day. Therefore, the two species are out of natural sync because they are governed by two different cues—one day length, the other temperature. When the bird migrates, its food source has already peaked, forcing the bird to nest faster in order to have something to eat. Dr. Root believes that eventually temperatures will rise high enough that it will leave the bird with little to eat once it has migrated because the caterpillars will already by gone.

This is especially pertinent to the natural world because as species adjust separately to global warming, natural established communities will have to adjust in order to survive. If adjustments take them out of sync concerning predator-prey species in the food chain, it will cause ecosystems to weaken and fail.

Dr. Root points out another serious issue connected to global warming: shifts in range. "Because habitats are quite fragmented these days, species will probably have a lot of difficulty shifting their ranges. It will be harder for species to cross farm fields, freeways, or cities. So if there is no other way for a species to shift, it could go extinct if it cannot adapt to a hotter climate. Adaptation is certainly possible, but not on the time scale that we are talking about here. Also, if the predator can shift and the prey cannot, there could be problems for the predator to feed itself."

According to Steve Schneider, a climatologist at Stanford University, "The best way to help species that are threatened by climate change is to stop using the atmosphere as a free sewer. This is a global problem and must be solved at a global level. The Kyoto Protocol [a voluntary international agreement designed to reduce greenhouse gas emissions] is not perfect, but it's a first step. It is embarrassing that the country that has poured more CO_2 into the atmosphere than any other countries put together [the United States]—even those with very large populations (China, for example)—is not doing anything about it. Each of us can do our part—drive hybrid cars; use fluorescent bulbs; when replacing furnaces, water heaters, appliances, windows, and roofs, replace them with the most energy efficient alternatives. But we need to put pressure on our policy makers to start negotiations on a global level about global warming."

Dr. Root also says, "A study has shown that without action to halt global warming, economists predict that the world as a whole will be ten times as rich by 2100, and people on average will be five times as well off. Adding on the costs of tackling warming would postpone this target by a mere two years. A two-year lag seems to me to be a pretty inexpensive planetary insurance policy!

"We need to continue to look for changes in the behaviors and cycles of animals and plants. It would be nice to have some studies that look at elevational transects up mountains and at different latitudes. I am currently looking at migration data and trying to understand how climate—primarily wind—has and could influence migration times."

Several species have already been identified as being in peril as a result of global warming. The polar bear is one of those. According to the Sierra Club, a well-established grassroots environmental organization whose goal is to protect communities, wild places, and the environment, the Arctic is on the front lines of global warming. Warming at a rate twice as fast as the rest of the Earth, climate change poses an extreme danger to the Arctic—a place where wildlife and nature have formed a delicate balance. So tenuous is the ecosystem that even small changes in temperature can cause devastating effects to the life within it. The IPCC, which also recognizes the polar region's unique vulnerability, warns that the severity of changes to habitat and its detrimental impacts to wildlife that are occurring

A polar bear mother and two cubs on the Beaufort Sea coast in the Arctic National Wildlife Refuge *(U.S. Fish and Wildlife Service)*

today are an early warning of the impacts that other wildlife will also face. They predict that other polar species such as walrus, seals, and penguins may soon find themselves in the same precarious situation as the polar bear.

Studies by the U.S. Department of Fish and Wildlife Services (FWS) indicate that polar bears are currently imperiled because of a warming Arctic climate. Studies by the FWS have documented plunging survival rates for cubs, falling body weights for adults, strandings on land for bears that are used to hunting for prey on vast expanses of ice, and even incidences of drowning in the oceans where ample, thick sea ice used to be. A report published by the U.S. Geological Survey (USGS) in 2008 indicates that America's polar bear populations could completely disappear by 2050 under these progressing conditions.

This threat is so great that the polar bear has now been added as "threatened" under the Endangered Species Act, the first time the U.S. government has added a species to the list because of the negative effects of global warming.

THE ENDANGERED SPECIES ACT

Half of the recorded extinctions of mammals over the past 2,000 years have occurred in the most recent 50-year period. The Endangered Species Act is a law that was passed by Congress and signed by President Nixon in 1973. It is regarded as one of the most comprehensive wildlife conservation laws in the world. In 1973, 109 species were listed. Today, more than 1,000 species are listed.

Congress had enacted two similar laws—one in 1966 and another in 1969—but neither did more than create lists of vanishing wildlife species. This was no better than publishing a list of murder victims but doing nothing to catch the murderers. The Endangered Species Act in 1973, however, changed all that. It forbids people from trapping, harming, harassing, poisoning, wounding, capturing, hunting, collecting, importing, exporting, or in any other way harming any species of animal or plant whose continued existence is threatened. The U.S. Fish and Wildlife Service manages the listing of land and freshwater species. The Natural Marine Fisheries Service is in charge of ocean species.

This act protects all species classified as threatened or endangered, whether they have commercial value (people want to buy, sell, and collect them) or not. It also forbids federal participation in projects that jeopardize listed species and calls for the protection of all habitat critical to listed species.

Under this law, "endangered" designates a species in danger of extinction throughout all or a significant part of its range. "Threatened" refers to species likely to become endangered in the foreseeable future. The Fish and Wildlife Service also maintains a list of "candidate" species. These are species that have not yet been proposed for listing but which the service has enough information about to warrant proposing them for listing as endangered or threatened. The protection also extends to the preservation of the species' habitats. Assistance is provided to states and foreign governments to assure this protection.

The law's ultimate goal is to "recover" species so they no longer need protection under the Endangered Species Act. The law provides for recovery plans to be developed and describes the steps needed to restore a species to health. Public and private agencies, institutions, and scientists assist in the development and implementation of recovery plans.

(continues)

(continues)

The Endangered Species Act is the law that puts into effect U.S. participation in the Convention on International Trade in Endangered Species of Wild Fauna and Flora (CITES)—a 130-nation agreement designed to prevent species from becoming endangered or extinct because of international trade.

There are some exceptions to the protection provided by the Endangered Species Act. Natives of Alaska can hunt endangered animals for use as food or shelter. An endangered animal can also be killed if it threatens the life of a human or if it is too sick to survive.

In the Arctic, permafrost that used to last 200 days during the year 50 years ago now lasts only 80 days each year. As global warming causes habitats to shift, animals will be forced to adapt, move, or die. Defenders of Wildlife, an organization dedicated to preserving wildlife and their habitats, states that more than a million species could go extinct in the next 50 years if action is not taken now to stop global warming and protect wildlife.

Penguins are also facing difficult challenges as icebergs melt. In 2004, a 1,200-square mile (3,108-km^2) iceberg trapped 3,000 nesting pairs of Adélie penguins on Cape Royds in Antarctica. The females, which normally had only a 1.25-mile (2-km) trip to the ocean for food now had to travel 112 miles (180 km) instead. Nearby, 50,000 penguin pairs on Cape Bird had to make a 60-mile (97-km) trip. These long excursions put all the penguins in peril; while the mothers are in search for food, the fathers must fast while they remain with the babies. If they have to fast for too long, they will abandon the chicks.

The waterfowl population of the prairie pothole region of the central United States and Canada is also at risk. Although this area accounts for only 10 percent of North American waterfowl breeding habitat, it is where 50 to 80 percent of all the continental duck production takes

Brown and white pelicans rest on Pelican Island National Wildlife Reserve. *(George Gentry. U.S. Fish and Wildlife Service)*

place. According to the Defenders of Wildlife, climate models predict that half of the pothole ponds will disappear in the next 100 years. The ponds that remain will support only 40–50 percent of the bird population they do now. Under drought conditions, in addition to loss of habitat, the ducks will have less nesting success, increased infant mortality, and make fewer nesting attempts.

Desert bighorn sheep herds that inhabit the Sonoran, Mojave, and Great Basin deserts of the southwest United States are dwindling in number. Less than 100 exist, and they are dispersed across the entire range in sparsely vegetated areas. Reports by Defenders of Wildlife indicate that western temperatures have recently exceeded the 100-year average by 2–3°F (1.2–1.8°C). In addition, California's precipitation has declined 20 percent. Because of this, plants and water that the sheep rely on for survival are disappearing. A recent inventory indicates that only 30 of California's 60 bighorn groups remain. Those that are left face extinction because of habitat loss and genetic diversity.

WORKING TOGETHER TO PROTECT WILDLIFE

Global warming is a problem that can be solved if all people work together. The following list provides some ways to work together to support global solutions to a global problem:

- Make contact with elected officials: Request that they support research and legislation that is designed to improve fuel efficiency standards and strengthens land manager and owners' ability to protect wildlife threatened by global warming.
- Retire your car: At least leave it home one day a week. Every gallon of gas saved keeps 20 pounds of CO_2 out of the atmosphere.
- Buy good wood: Use Forest Stewardship Council (FSC) logos on wood products. FSC makes sure trees are grown and harvested in a way that protects wildlife and soil quality.
- Buy organic and buy local: Chemical fertilizers and insecticides, as well as transportation and production costs, require the use of excessive amounts of energy.
- Coordinate an activist event: Visit wildlife refuges, zoos, and parks. Get together with friends to discuss global warming, or even watch a movie such as *An Inconvenient Truth.*

There are several ways global warming is affecting various species of wildlife, which is why the National Wildlife Federation considers global warming to be "the most dangerous threat to the future of wildlife."

There are several species in peril as temperatures rise in the Arctic. Mosquitoes multiply at a rapid rate and expand their habitat farther northward. As the caribou spend more time and energy evading the pests, they are decreasing the amount of food that they eat and energy they conserve in preparation for the coming winter months. This particularly puts the females at risk, taking the energy from them necessary to raise their young.

- Write a letter to the editor of your local newspaper: Raise awareness in the media and the public by highlighting global warming, wildlife, and habitat in your local paper.
- Volunteer for wildlife: Participate in volunteer activities with local, state, or federal land and wildlife management agencies, universities, and other groups.
- Reach out to your community: Allow organizations to assist in taking a stand. Take a wildlife protection message to organizations and events in the community.
- Welcome wildlife into your garden: Create habitat by planting native plants.
- Light bulbs, thermostats, and washing machines matter: Reduce annual CO_2 emissions by using compact fluorescent bulbs, cold water to wash clothes, and lower your heating thermostats two-degrees.
- Speak out for better fuel economy standards: The United States could cut emissions by 600 million tons per year if vehicles averaged 40 miles per gallon.
- Buy Green Power: Check DOE's green pricing page to locate local options of purchasing green power.

Source: Defenders of Wildlife

The monarch butterfly is another species in peril. In the summer, they migrate as far north as Canada, in the winter as far south as Mexico City. In 2002, there was a mass die-off of these butterflies because of a change in climate. Scientists worry that as temperatures and precipitation patterns continue to change, more die-offs will occur. Migratory songbirds, which are also extremely sensitive to changes in temperature, may also be in peril as their habitats change.

North American trout and other coldwater fish depend on extremely cold spring meltwater for their survival, but as temperatures continue to rise in North America, 75 percent of their natural habitat is at risk. Coral reefs do not fare much better. These colorful underwater ecosystems are extremely fragile. As ocean temperatures climb, the corals become bleached and die off. The die-off is not specific to one location, either. In fact, nearly 16 percent of the world's reefs were destroyed in less than a 12-month period. The increase in temperature does not even need to be drastic; an increase of 1.7°F (1°C) is enough to cause coral die-off.

According to the National Wildlife Federation, high-mountain species are particularly vulnerable to the effects of global warming. As climate conditions change and habitats move upward in elevation, those species living in habitats already at the highest elevations are pushed upward with no place to go. The pika is one example of this found in the North American Rocky Mountains. Pikas are potato-size herbivores that inhabit the mountaintops. A member of the rabbit family, they live their entire lives in the alpine regions of mountaintops. A hearty species, able to survive the most severe weather conditions during the intense Rocky Mountain winters, biologists such as Chris Ray at the University of Colorado fear that these resilient creatures may not survive global warming. Unlike other species that are currently shifting their ranges north in latitude or higher in elevation on mountains in response to climate change, the pika have nowhere else to go. In fact, according to Ray, in some areas entire pika populations have already disappeared.

In Rocky Mountain National Park, which has more than 100 square miles (259 km^2) of alpine habitat, researchers at Colorado State University estimate that the tree line there would rise 1,200 feet (366 m), eliminating half of the park's tundra if temperatures warm by 5°F (3°C). Even now, trees are already migrating to higher elevations. A study published in the *Western North American Naturalist* in July 2005 determined that the low-elevation Engelmann spruce, which lives in the subalpine zone, has already migrated 575–650 feet (175–198 m) upslope in three of four watersheds that were studied in Nevada's Great Basin National Park between 1992 and 2001.

The alpine wildlife that inhabits the highest reaches is especially vulnerable to the disastrous effects of global warming. They are especially vulnerable to changes in vegetation, the invasion of new predators and pests, reduced winter snowpack, and increases in severe weather. In the case of pikas, they are especially vulnerable to excessive heat.

Accustomed to cool temperatures, they inhabit the cool, wet talus (rock rubble) at the bases of mountains during the summer. As the climate warms, the pika are expected to move upward in elevation, looking for cooler places to stay, until there is nowhere else left to ascend.

According to Barry Rosenbaum, a Colorado College alpine mammalogist, "All other mammal species in continental North America have greater heat tolerances." He is currently studying the pika on Colorado's Niwot Ridge.

Erik Beever, a biologist with the National Park Service, is studying the pikas in the Great Basin region—the arid region between the Rocky Mountains and the Sierra Nevada. He says, "The mammals have recently disappeared from 8 of 25 mountainous locations where they were documented in the early 1900s." He says that the die-off indicates

The pika is one animal that is already feeling the negative effects of global warming: Its habitat is being eradicated at the mountaintops as other species migrate in. *(U.S. Fish and Wildlife Service)*

that suitable habitat is shrinking. To support the global warming as a cause, the most recent pika losses occurred at the warmer, southern end of the animal's range.

Beever remarks, "This is what you would expect from rising temperatures—a loss at the margins of their distribution. The finding represents one of the first contemporary examples of a North American mammal exhibiting a rapid shift in distribution due to climate."

Chris Ray refers to the situation as "the proverbial canary in a coal mine. It is a warning that global warming is upon us now."

THE TIMING OF SEASONAL EVENTS

The loggerhead sea turtle is another species being documented as subjected to the negative effects of global warming. John Weishampel, an associate professor at the University of Central Florida, has analyzed long-term records of a 25-mile (40-km) length of Florida coast, which represents one of North America's key breeding grounds for the loggerheads and has discovered that the turtles today are coming ashore to lay their eggs about 10 days earlier than they were in 1989. Suspecting global warming as the cause for the change in breeding behavior over the past 20 years, he analyzed water temperature data that was collected offshore and found that temperatures had increased by 1.5°F (0.9°C) over this time period.

Weishampel's findings are consistent with findings of other scientists worldwide who are studying different ecosystems for the same reasons. Based on a study published in *Global Change Biology* in 2007, as global temperatures continue to rise, more discoveries are made indicating that the timing of seasonal events in the life cycles of both plants and animals are making dramatic changes—shifting out of sync with longtime patterns.

Trees worldwide are budding earlier in the spring and losing leaves later in the fall because rising temperatures are causing spring warm-up to occur earlier and summerlike temperatures to linger later in the fall. Animals are also exhibiting natural shifts by migrating, mating, producing young, and emerging from hibernation earlier in the spring. Other animals have also been similarly affected. Based on a report from the Pew Center for Global Climate Change in November 2004, more than three dozen scientific reports have linked global warming to ecological changes in the United States. The shifts are consistent across species, ecosystems, and geographic regions throughout the country. The following list summarizes the findings of the reports:

- Plants in Washington, D.C., are flowering 4.5 days earlier than they did 30 years ago.
- Mexican jays in southern Arizona breed about 10 days sooner than they did in 1971.
- Barn swallows nationwide are nesting nine days sooner than they did in 1959.
- In Michigan's Upper Peninsula, birds are returning from their winter migrations earlier in the spring.
- The Eastern Seaboard's Cape May warbler has shifted its range to the north, and if the trend continues, in a century there will be no more Baltimore orioles in Baltimore.
- American robins are arriving to breed two weeks earlier than they did in the late 1970s in response to warmer temperatures in their low-latitude winter habitat. Currently, when the birds arrive, it is still winter. They must wait for the snow to melt before they can feed themselves and their young.

"The yellow-bellied marmots are also responding to higher spring air temperatures by leaving hibernation dens more than a month sooner than they did a few decades ago. But if the mismatch between cues at lower and higher elevations keeps growing, these animals could be in trouble," says biologist David Inouye, of the University of Maryland.

In each case, the species' behavioral change coincides with a temperature increase during the same period.

Camille Parmesan, a biologist at the University of Texas, says, "Climate change is happening right here, right now, and changing life in your own backyard."

Parmesan also says that a major concern is that "any given species is unlikely to respond to changing climate in the same way as others that share its habitat. When this happens, vital links among interdependent species—such as a plant and its pollinator—can be broken, a loss of

THE MYTHS AND REALITIES ABOUT THE ENDANGERED SPECIES ACT

The U.S. Fish and Wildlife Service has identified common myths about endangered species. For example:

Myth: Some people believe that extinction is a "natural" process and that we should not worry about it.
Reality: The reality is that while extinction in general may be a normal process, the high extinction rates that exist today are not. In many cases, the environment is changing so fast that species do not have time to adapt. Scientists have determined that a natural extinction rate is one species lost every 100 years. However, since the Pilgrims landed at Plymouth Rock over 365 years ago, more than 500 North American species have become extinct—that is more than one species becoming extinct per year.

Myth: The Endangered Species Act is causing loss of jobs and economic strain in many areas of the country.
Reality: When economists from the Massachusetts Institute of Technology analyzed the economic impact of endangered species, they found that states with many listed species have economies just as strong as those areas that do not have endangered species.

Myth: Billions of tax dollars are being spent on endangered species.

synchrony that not only threatens each partner in the relationship, but may also disrupt entire plant and animal communities."

John Weishample notes, "In nature, timing is everything. It's like a symphony that's ruined if one instrument comes in at the wrong time."

Doug Inkley, senior scientist of the National Wildlife Federation (an organization whose goal is to protect global wildlife diversity), says, "Global warming presents a profound threat to wildlife in this country, a threat that may equal or even surpass the effects of habitat destruction."

Inkley refers to a report that appeared in both *Global Climate Change* and *Wildlife in North America,* published by the Wildlife Society,

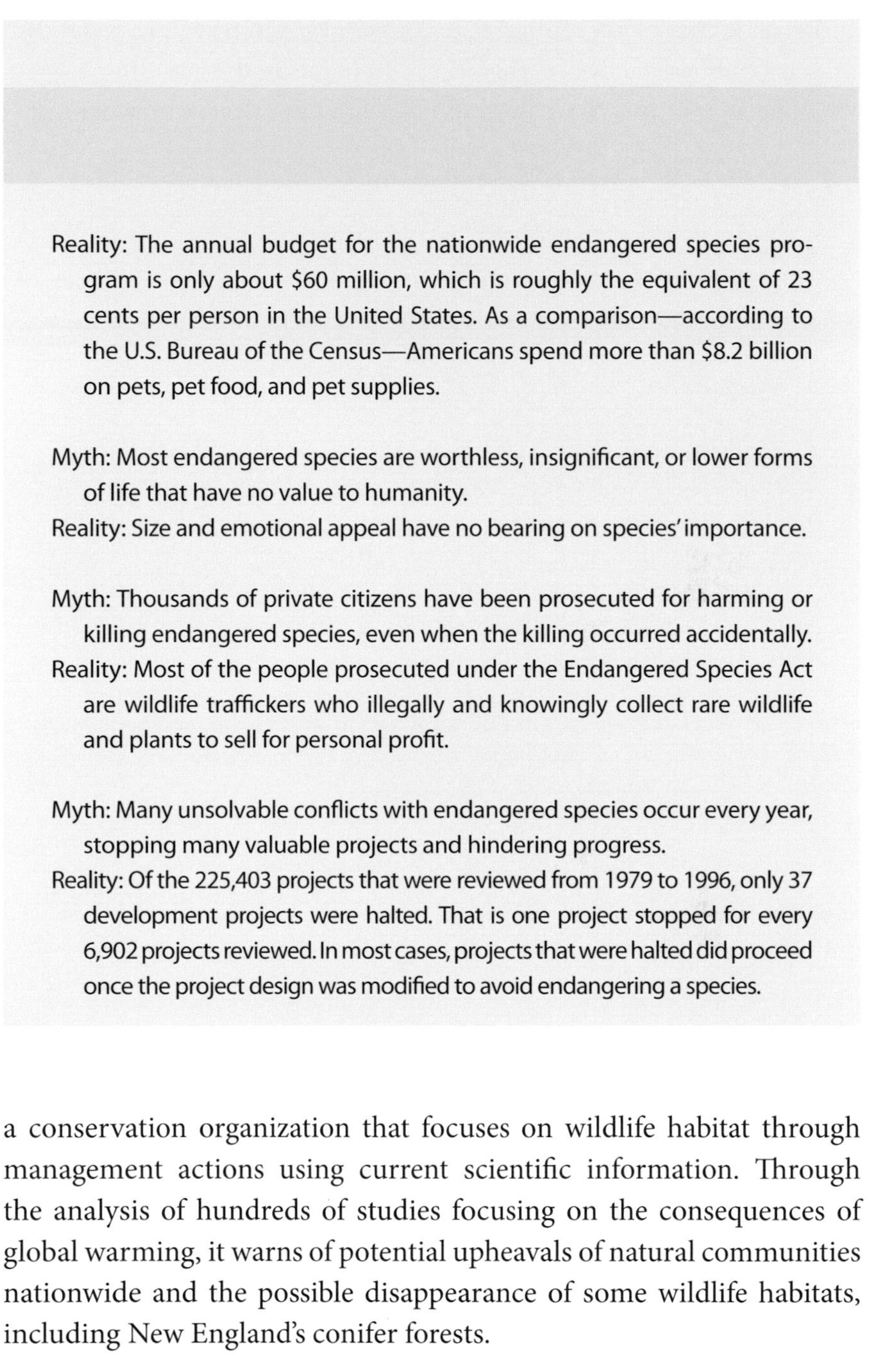
Reality: The annual budget for the nationwide endangered species program is only about $60 million, which is roughly the equivalent of 23 cents per person in the United States. As a comparison—according to the U.S. Bureau of the Census—Americans spend more than $8.2 billion on pets, pet food, and pet supplies.

Myth: Most endangered species are worthless, insignificant, or lower forms of life that have no value to humanity.
Reality: Size and emotional appeal have no bearing on species' importance.

Myth: Thousands of private citizens have been prosecuted for harming or killing endangered species, even when the killing occurred accidentally.
Reality: Most of the people prosecuted under the Endangered Species Act are wildlife traffickers who illegally and knowingly collect rare wildlife and plants to sell for personal profit.

Myth: Many unsolvable conflicts with endangered species occur every year, stopping many valuable projects and hindering progress.
Reality: Of the 225,403 projects that were reviewed from 1979 to 1996, only 37 development projects were halted. That is one project stopped for every 6,902 projects reviewed. In most cases, projects that were halted did proceed once the project design was modified to avoid endangering a species.

a conservation organization that focuses on wildlife habitat through management actions using current scientific information. Through the analysis of hundreds of studies focusing on the consequences of global warming, it warns of potential upheavals of natural communities nationwide and the possible disappearance of some wildlife habitats, including New England's conifer forests.

Peter Marra, an ornithologist at the Smithsonian Environmental Research Center, says that for every 1.8°F (1°C) increase in springtime temperature at the breeding sites, migrants are arriving one day earlier, probably because they are able to pick up clues en route—such as earlier greening or more insects to eat—and begin traveling faster. But migrants still were not keeping pace with plants in the breeding habitat, which were budding three days sooner with each degree of warming.

WILDLIFE IN PERIL

Changes in the timing of seasonal behavior are one way that global warming is affecting wildlife. It is also causing species to migrate poleward in latitude and higher in elevation in mountain ecosystems. Some of the species having the most difficult time adjusting to changes in temperature and what that means for their habitat are some of Earth's most vulnerable and endangered wildlife species. The following list depicts some of the species that are most threatened:

- *Black Guillemots*: On Cooper Island in Alaska, they are declining as the sea ice melts, taking away the bird's primary food source, Arctic cod (which live under the ice).
- *Sockeye Salmon*: Abnormally high water temperatures and drought-induced low water flows have already killed tens of thousands of sockeye salmon in British Columbia. Scientists warn there will be more die-offs as temperatures rise.
- *Sooty Shearwaters*: These have decreased by 90 percent on the U.S. West Coast due to changes in ocean temperatures and currents as a result of global warming.
- *Snails, Sea Stars and Other Intertidal Creatures*: In Monterey Bay, California, they have been migrating northward for the past 60 years as sea and air temperatures get warmer.
- *Edith's Checkerspot Butterflies*: These extend from the west coast of southern Canada through northern Mexico, but their range is shrinking. In the southern reaches of their habitat, they have already become extinct.
- *Mexican Jays*: In southern Arizona, the breeding season has advanced 10 days between 1971 and 1998. During this time, spring temperatures have consistently risen 4.5°F (2.7°C).
- *American Robins*: In Colorado, they are migrating from low to high elevations where they now breed two full weeks earlier than they did in the late 1970s.

Marra also noted, "There's probably a limit to the birds' flexibility. Migration birds evolved to cope with gradual climatic changes, but the changes today are happening much faster."

- *American Pikas*: In the U.S. Great Basin, eight of the 25 major populations no longer exist. Already occupying the habitats highest in elevation on the mountaintops, they have nowhere to migrate when temperatures rise.
- *Polar Bears*: Melting sea ice has decreased the amount of time polar bears have to hunt seals, leaving them with a shortage of food. As a result, polar bears weigh less and have fewer cubs than they did 20 years ago.
- *Red-Winged Blackbirds*: The birds arrive at their breeding grounds in northern Michigan three weeks earlier than they did in 1960.
- *American Lobsters*: In western Long Island Sound, large numbers of lobsters died mysteriously during September 1999. Some researchers believe it was the result of warmer water.
- *Prothonotary Warblers*: These travel from South America to Virginia a day earlier each year as springtime temperatures rise.
- *American Alligators*: Their distribution, which ranges from the Carolinas south to Florida and west to Texas, is shifting northward in some regions. Rising sea levels may also push them farther inland where they will interfere with developed areas.
- *Loggerhead Sea Turtles*: Loggerheads are threatened and need protection. Because of warming, they are coming ashore to nest 10 days earlier than they did in 1989.
- *Coral Reefs*: High water temperatures are causing corals to expel their symbiotic algae—or "bleach"—which can lead to coral death and damage to entire reef ecosystems.
- *Golden Toads*: This is the first species whose extinction was caused by global warming.

Global warming affects a wide diversity of species worldwide, which is the major reason why the solution to this problem needs to be *global* in scope.

Source: National Wildlife Federation

Beyond shifts at the species level, entire communities are being transformed as plants and animals in the same habitat respond differently to climate change. According to Terry Root, "In the future, well-balanced wildlife communities as we know them will likely be torn apart."

REFUGES AT RISK

Global warming is the single greatest threat imperiling the National Wildlife Refuge system. As climate changes, it will cause changes to ecosystems and habitats and reduce their ability to adequately support wildlife. Defenders of Wildlife (an organization dedicated to the preservation of all wild animals and native plants in their natural communities) has determined there will be major changes to plant communities and for storm intensity, sea levels, and water supplies. These issues will affect the natural refuges in place today, limiting their future ability to support wildlife.

Currently, more than 160 National Wildlife Refuges in the United States exist along the 95,000 miles (152,887.7 km) of U.S. coastline. They are in serious danger of being affected by global warming because scientists expect sea levels to rise as polar ice and glaciers melt and oceans physically expand as they get warmer. It is expected that sea-level rise will range from four to 36 inches (10–91 cm). Negative effects to these refuges include: subsidence and erosion of the land; saltwater contamination of freshwater; inland shifts of beaches; loss of coastal wetlands and mangroves; declines in coral reef health; changes of forest habitat into marsh; and increases in coastal areas' susceptibility to frequent, aggressive storms.

The prairie region in the United States is another area of concern. This area is the most critical waterfowl-producing region in North America, with its "prairie potholes," featuring extensive grasslands and large wetlandlike depressions. The potholes are critical to avian migration.

The Environmental Protection Agency (EPA) acknowledges that as temperatures rise, soil moisture can decrease, especially during the summer. The agency predicts that warmer climates in the northern prairie wetlands region will increase the frequency and severity of droughts, which could reduce the potholes from 1.3 million to 800,000 by 2050. In addition, because waterfowl breeding habitat is so concentrated in the pothole region, the effects of global warming could potentially cut the number of breeding ducks in half and increase the likelihood of avian flu and other deaths.

The wildlife refuges in Alaska are also under close watch. According to Defenders of Wildlife, since 1950, the average temperature in the Arctic region has increased by 4–7°F (2.4–4.2°C). Precipitation has increased by 30 percent since 1968. Evidence of warming is everywhere—melting sea ice, melting glaciers, thawing permafrost, migrating wildlife, and insect infestations.

Research completed by the U.S. Global Change Research Program projects that Alaska will warm 1.5–5°F (0.9–3°C) by 2030 and 5–18°F

THE ROLE OF REFUGES

All over North America, wildlife that was once shot or trapped year-round is now protected for at least part of each year by state and/or federal laws. Historically, it was overhunting that led to the establishment of the first federal wildlife refuges in the United States. The refuge system began because all wildlife—including game birds such as ducks and other migratory waterfowl—were being slaughtered. President Theodore Roosevelt realized that waterfowl needed safe places along their migration routes. By 1904, 51 refuges had been set aside in the United States and its territories. Many were mainly waterfowl habitats, but some were established for the benefit of nongame birds such as pelicans and spoonbills.

Today there are more than 440 federal wildlife refuges totaling 92 million acres. Most of the acreage is in Alaska, but refuges are located all over the United States and its territories. Also, many national parks, forests, and monuments; thousands of state and county parks and reserves; and many privately owned preserves provide habitat for wildlife in the United States. Millions of acres have been set aside for the protection of wild animals and plants. Many environmental groups, ecologists, and private citizens would like to see more land set aside for these efforts. It often takes many years of public education and political lobbying in order to establish new reserves to protect wildlife. The effort to promote species diversity is well rewarded.

(3–10.8°C) by 2100. Northern Alaska is expected to experience the most extreme of the effects, particularly in the winter. The researchers also project that the ground will continue to thaw and that the upper 30–35 feet (9–11 m) of permafrost will probably melt by 2100. In addition, precipitation will increase up to 25 percent in the north and northwest, but decrease by 10 percent along the southern coast. Soils will dry out due to increased evaporation, and wildfires will become more prevalent. As these changes take place, the effects on the refuges will be devastating to the plant and animal species living within them.

A report by Defenders of Wildlife released on October 4, 2007, identified the 10 refuges most at risk in the United States. Issues such as global warming, lack of funds, oil and gas development, road construction, invasive species, and water contamination are all playing a role putting these refuges at risk. They are:

- Cape May National Wildlife Refuge, New Jersey
- Hailstone National Wildlife Refuge, Montana
- Lower Rio Grande Valley National Wildlife Refuge, Texas
- Nisqually National Wildlife Refuge, Washington
- Pea Island National Wildlife Refuge, North Carolina
- Rappahannock River Valley National Wildlife Refuge, Virginia
- Rhode Island National Wildlife Refuge Complex, Rhode Island
- San Luis National Wildlife Refuge, California
- Trempealeau National Wildlife Refuge, Wisconsin
- Yukon Flats National Wildlife Refuge, Alaska

Cape May National Wildlife Refuge is an estuarine ecosystem that contains the world's largest spawning ground for horseshoe crabs. Because the horseshoe crabs lay their eggs there, thousands of migratory songbirds, such as the ruby-crowned kinglet and the Nashville warbler arrive at the refuge, which serves as a critical stopping ground they use along the Atlantic Flyway. The refuge provides a habitat for 100 peregrine falcons, 7,000 American kestrels, and 150 northern harriers. Up to 317 bird species can be seen at the refuge, including the largest concentration of American woodcocks on the Atlantic coast.

The refuge also provides a home to a multitude of mammals, reptiles, amphibians, fish, shellfish, and invertebrates. Cape May currently protects 21,000 acres (8,498 hectares) of wetlands, forests, grassland, shoreline, and salt marsh. If global warming continues and habitats

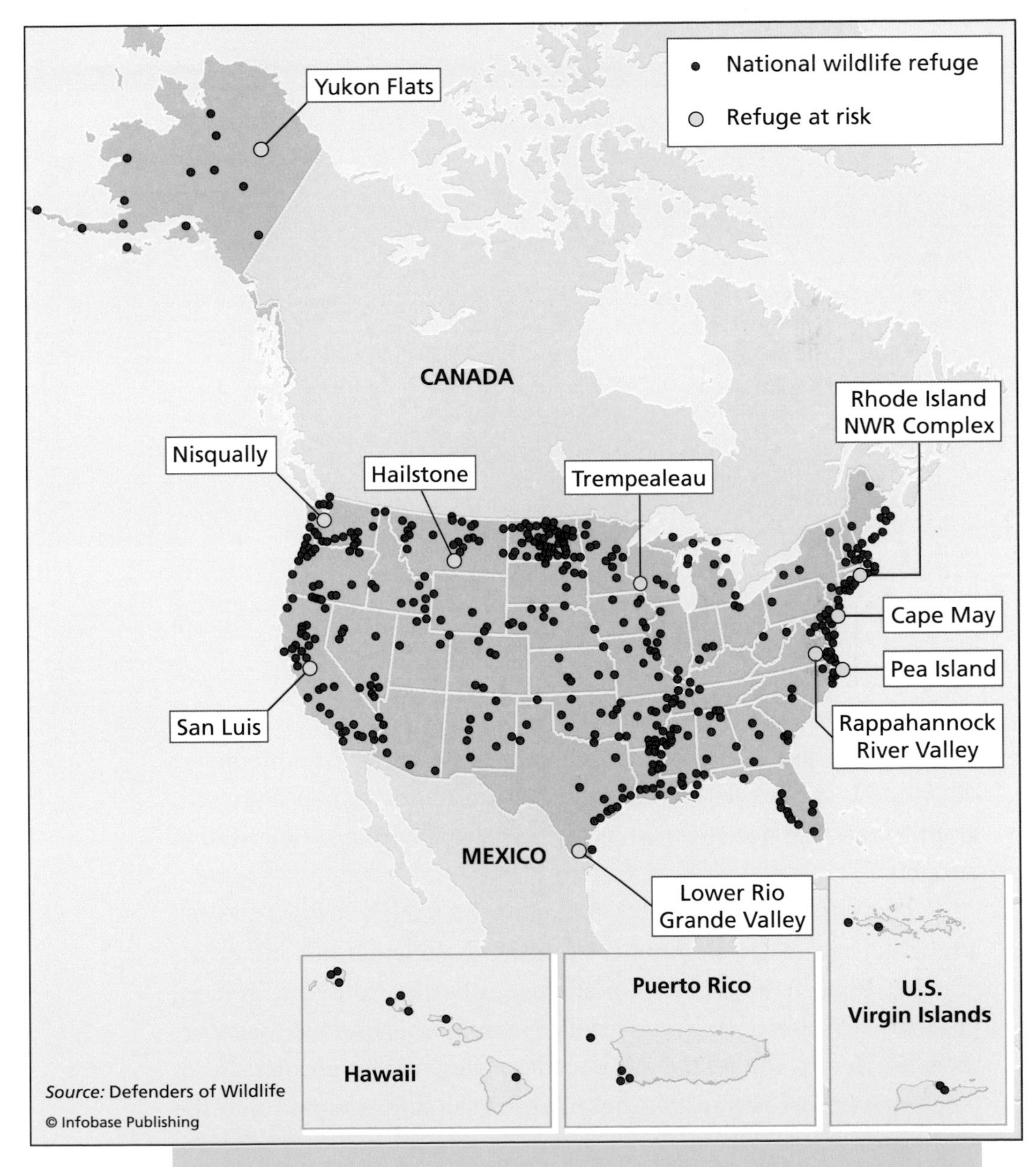

The 10 refuges in the United States most at risk *(Defenders of Wildlife)*

The endangered Florida panther—the panther is currently protected at the Florida Panther National Wildlife Refuge. *(U.S. Fish and Wildlife Service)*

shift, the ecosystems within the refuge may no longer be suitable for the wildlife species within, forcing them to migrate to another geographic area, but no longer be under the protected status of a wildlife refuge.

Hailstone National Wildlife Refuge in Montana provides a home and major migratory stopover for birds. The wetlands and shortgrass prairies support thousands of mallards, gadwalls, American avocets, phalaropes, American white pelicans, Franklin's gulls, and vesper sparrows. Currently, the refuge has become polluted from an increase in precipitation that is transporting dissolved salt and selenium into the soil, which enters the water table and flows into the reservoir, where it has become toxic to the wildlife at the refuge. If global warming results in increased precipitation to the area, the health of the wetlands will continue to disintegrate and poison the wildlife within it.

The Lower Rio Grande Valley National Wildlife Refuge in Texas offers habitat to two species of endangered cats—the ocelot and the jaguarrundi. The refuge lies in the northernmost range of their natural habitat. The refuge encompasses 90,000 acres (36,422 hectares) but is fragmented into 115 parcels. While the refuge provides the last of the remaining habitat in the United States to these two feline species, it also provides habitat for 513 bird species, some of which are rarely seen anywhere else in the United States. These species include the least grebe, green parakeet, altamira, oriole, green jay, and the great kiskadee. The refuge is unique because it covers four different climate regimes: tropical, temperate, coastal, and desert. Considered one of the most diverse regions in North America, it encompasses 11 distinct biotic communities. It contains the Chihuahuan desert scrub, tidal wetlands, and one of the last remaining sabal palm forests in the United States. Because of the diverse spread in natural communities, it provides a habitat for more than 1,100 plant species, 700 vertebrates such as javalinas, bobcats, white-lipped frogs, and the endangered Kemp's Ridley sea turtle. Within its boundaries live more than 300 different species of butterflies, which represents half of the total species found in the United States. If global warming persists, the unique habitats will face degradation as the climatic zones degrade or shift northward, causing the wildlife to shift toward unprotected areas that may not be able to support their needs for survival.

The Nisqually National Wildlife Refuge in Washington at Puget Sound is where the Chinook salmon pass en route to the waters northward of their birth. Also present at the refuge are thousands of western sandpipers, dunlin and other shorebirds that forage mudflats for worms, clams, crab, and shrimp. Pacific tree frogs inhabit the area, along with harbor seals and river otters, and wrens reside in abundance in the estuary. The refuge has been designated as a National Natural Landmark, representing one of the state's last undisturbed estuary ecosystems.

Environmental conditions today at the refuge include a decline in forest health, a decline in freshwater wetland health, a surge of incoming invasive-plant migration, and deterioration in the quality of the

Nisqually watershed. Global warming in the future will continue to decrease the quality of habitat for resident wildlife.

Pea Island National Wildlife Refuge on the northern end of North Carolina's Hatteras Island provides a home for migrating birds and threatened loggerhead turtles. It also serves as a flyway for migrating birds such as the snow geese, ruddy turnstone, dunlin, and marbled godwit. Its bayside marshes and ponds are populated with herons, avocets, tundra swans, northern pintails, and cedar waxwings. Along its 13-mile (21-km) stretch of beach, the refuge provides a habitat for more than 365 species of birds. Located along the coast, it is in the pathway of severe storms, sea-level rise, and warmer ocean temperatures—all negative impacts from global warming.

Rappahannock River Valley National Wildlife Refuge in Virginia, which is bounded on its borders by the Rappahannock River to the south and west, the Potomac River to the north, and the Chesapeake Bay to the east, provides a habitat for a large population of bald eagles. With approximately 400 eagles inhabiting the refuge, it makes the Rappahannock a significant area for bald eagle conservation. In addition to the eagle, it also provides habitat for more than 225 species of other birds such as Indigo buntings, Acadian flycatchers, red-winged blackbirds, and eastern meadowlarks. Mammals, reptiles, and amphibians inhabit the grasslands and ponds. Refuge waters are habitat for bass, catfish, croaker, and the endangered shortnose sturgeon. The refuge is broken into several separate parcels, and as global warming continues to push climate zones to new areas, wildlife will have to be relocated as the fragmented lands evolve and invasive species encroach, rainfall patterns change, and temperatures vary.

The Rhode Island National Wildlife Refuge Complex in Rhode Island is home for a population of harlequin ducks, which share the Atlantic shoreline with multitudes of other wildlife such as harbor seals and woodcocks. It also provides shelter and home for migrating avian species such as raptors. More than 300 species of birds—snow buntings, eiders, scoters, loons, yellow-breasted chats, and American black ducks—occupy the refuge. Officially part of the Atlantic Flyway, it is one of the most important migratory habitats on the East Coast. Like other refuges located in coastal areas, it faces the same vulnerabilities to global warming.

The San Luis National Wildlife Refuge at the northern end of the San Joaquin Valley in California is a 130,000-acre (43,479-hectare) habitat that plays a key role for migrating birds along the Pacific Flyway. Each year, millions of migratory birds such as green-winged teal, ring-neck ducks, snow geese, and sandhill cranes use the refuge as a stopover on their long annual migration.

Most of the prime wetland area in California has already been drained, filled in, or destroyed; most of it has been converted to agriculture, making this refuge a key resource for wildlife—both migratory and local. The refuge contains more than 210 species of birds, including sandpipers, dunlins, black-necked stilts, herons, egrets, raptors, and songbirds. It also provides habitat for mammals, reptiles, and amphibians, including rare species such as the endangered San Joaquin kit fox and Tule elk (the smallest elk in North American and once nearly exterminated because of habitat loss and overhunting in the early 1900s).

The refuge is struggling today, however, in obtaining enough freshwater due to the excessive demands of nearby urbanization and agriculture. The refuge is currently having difficulty obtaining enough water to meet its needs. The larger concern is that it will not be able to compete on the open market for California's limited water resources, which are expected to become even scarcer as prolonged drought and global warming escalate.

Trempealeau National Wildlife Refuge in Wisconsin provides a refuge in its lush marshlands. Located at the confluence of the Trempealeau and Mississippi Rivers, it is habitat for black terns, wood ducks, monarch butterflies, blue-winged teal, and hooded mergansers. The refuge serves as a flyway for migratory birds. In addition to the marsh and open pools, however, there are also forests, meadows, and a sand prairie.

The refuge faces many challenges today, however, with rising temperatures and changing climates. Invasive plants are encroaching the refuge, such as leafy spurge, purple loosestrife, quackgrass, smooth bromegrass, and black locust trees. These invasive species are killing off the native hardwoods, grasses, and forbs. As global warming continues, more invasive species will threaten and destroy the indigenous species.

Yukon Flats National Wildlife Refuge in Alaska at the northernmost stretch of the Yukon River is home to millions of waterfowl such as canvasbacks, pintails, wigeon, scaup, and shovelers. Wildlife arrives from 11 different countries, eight Canadian provinces, and 45 of the U.S. states. Yukon Flats supports one of the highest nesting densities of waterfowl on the North American continent and has become an increasingly important breeding area as prairie pothole habitat in the lower 48 states and Canada have been degraded by agriculture, development, and global warming.

Yukon Flats also provides a habitat for black bears, grizzlies, moose, caribou, and lynx. One of the largest challenges the refuge faces today is that of oil exploration and development. Because of its extreme northern location, it is also a fragile area and may face peril as climate changes and temperatures continue to rise.

LAWS IN FORCE TO HELP WILDLIFE

At a federal land and wildlife conference held in Boise, Idaho, in June 2008, the focus was on how global warming will affect natural resources management. The take-home message delivered at the end was that "global warming and wildlife do not mix." Natural resource managers there reported that climate change threatens so many species today worldwide that it will be impossible to save them all. Today's wildlife refuges may end up too far south or too low in elevation to house the animals they were created to protect. National parks that the federal government strives to keep in their natural state may eventually face a far different climate than the one they had when they were created. The conference also concluded that wildlife managers and federal policy makers may have to make what is known as "Noah's choice," named after the biblical story of Noah, who saved animals from the flood. With so many species facing extinction and limited resources (such as funding), someone may have to choose which animals will survive and which ones will not.

The conference represents a significant stride for government leadership. It marks the first time that resource agencies have brought together the people who manage the refuges, parks, wildlife populations, forests, and rangeland, encouraging them to work together toward a common goal. The changing reality of the ecosystems they manage is now forcing them to reconsider their priorities and the way they will do their jobs in the future.

Anne Kinsinger, the western regional director of the U.S. Geological Survey, who organized the conference, says, "We realize that so much of what we know is looking backward. That's no longer relevant."

She stresses that currently managers are striving to maintain the ecosystems that have existed for hundreds of years. With global warming, however, maintaining those cooler ecosystems simply will not be possible.

Jeff Burgelt, a biologist with the U.S. Fish and Wildlife Service, said, "Managers are going to face challenges like none they have seen in their careers. As species and the ecological communities on which they depend move north and up mountains to adjust to warming temperatures, existing wildlife refuges will become obsolete and management plans worthless."

"We need to develop an entirely new conservation paradigm," said Jean Brennan, a climate scientist with Defenders of Wildlife. "Instead of protecting the habitat species' needs today, managers must begin to develop predictive tools to preserve their habitat in the future. They need to be able to make it possible for the animals and plants to move to the new areas in a landscape fragmented by human development."

The Safe Climate Act of 2007

The Safe Climate Act of 2007 (H.R. 1590), introduced by Representative Henry Waxman, Democrat from California, is a bill designed to take action to avoid the predicted, most severe effects of global warming on the environment. The bill focuses on increasing reliance on renewable energy sources and efficient energy. It establishes pollution reduction targets that aim to keep temperatures below the 2°F (1.2°C) "tipping point" that scientists refer to as the dangerous point beyond which drastic climate changes will be unavoidable. In order to stay below the tipping point, the concentrations of global warming pollutants must not exceed 450 ppm by 2100.

The bill establishes pollution reduction targets designed to keep levels below this limit. Emission levels are set at 2010 levels and then gradually reduced through 2050. From 2011 to 2020, emissions are cut 2 percent per year, after which reductions are cut 5 percent per year.

Under these guidelines, by midcentury, emissions are calculated to be 80 percent lower than in 1990.

In order to reach these goals, the bill requires an increased reliance on clean, renewable energy sources, improved energy efficiency, and clean cars. It directs the Department of Energy (DOE) to establish national standards requiring 20 percent of electricity to be generated from renewable energy resources by 2020. It also requires utilities to obtain 10 percent of their energy supplies through energy efficiency in 2020.

Standards must also be set for reducing global warming emissions from motor vehicles, which are at least as rigorous as the current California standards. It also directs the EPA to set a cap on global warming emissions from the largest polluters and allow polluters to meet the cap by buying and selling pollution allowances—a type of "pollution permit." The funds that are generated from the allowances sales are then deposited in the "Climate Reinvestment Fund," and are then used for public benefit and to promote economic growth, including helping fish and wildlife adapt to a changing climate.

The bill also requires the National Academy of Sciences to review on a five-year cycle the progress in order to enable adjustments to be made where necessary.

The Global Warming Wildlife Survival Act

The Global Warming Wildlife Survival Act provides a national approach to manage the impacts of climate change on wildlife. It was introduced by Representatives Norm Dicks, (D-WA), Jay Inslee, (D-WA), and James Saxton, (R-N.J.), and was included in H.R. 2337—a comprehensive energy and global warming bill sponsored by Rep. Nick Rahall, (D-W.V.), who chairs the House Natural Resources Committee. The legislation was included in the multicommittee New Direction for Energy Independence Act (H.R. 3221), passed by the House of Representatives on August 4, 2007, by a vote of 241 to 172.

The purpose of this act is to prepare for the current and future effects of global warming on ecosystems and wildlife. Because of the longevity of greenhouse gases in the atmosphere, the negative impacts to ecosystems and wildlife will persist for at least the next 100 years because of what has already been put into the atmosphere.

The major goals of the Survival Act include:

- Ensuring that federal and state government agencies develop sound plans to reduce the effects of global warming on wildlife and its habitat.
- Establishing a national scientific advisory council to determine the impacts of global warming on wildlife.
- Developing and coordinating a national plan to provide a way for wildlife to adapt to both the current and future impacts that global warming will have on their environment over the next century as plans are put into action to reduce emissions from energy sources.
- To structure a federal-level plan of funding that will best serve the sustainable survival of wildlife during the global warming process.

One of the strongest points of this act is that it focuses on bringing management agencies together from multiple scales of jurisdiction: local, state, and federal government, as well as private entities. It is geared toward the management of land on a "whole-habitat" basis, not on one defined by political jurisdictional boundaries—a recurrent problem of past management regimes. Instead of developing different conservation and planning programs for individual tracts of land based on what agency manages it, its management will be conducted as a comprehensive ecological system. Because global warming is such a complex problem, future management plans and strategies will be developed through a coordinated national strategy to ensure that wildlife impacts spanning governmental jurisdictions are effectively addressed and appropriately funded. It will also focus on making sure that the necessary scientific expertise is available to make educated, effective, forward-looking decisions.

To date, each state has already completed a wildlife action plan that addresses global warming impacts on wildlife, has identified at-risk habitats and species that need special conservation attention, and coordinated these plans with a national strategy.

The next focus has been identified as creating a new national interagency global warming scientific support center in order to coordinate and focus scientific efforts. The center would have several objectives:

- research coordination;
- federal land management support;
- creation of a comprehensive scientific inventory;
- development of effective monitoring programs within each federal land management agency.

The Global Warming Wildlife Survival Act is also important because it provides the authorization for federal funding needed to implement the act.

According to Defenders of Wildlife: "Congress must pass meaningful, aggressive legislation to immediately reduce greenhouse gas emissions." This includes legislation that moves the United States away from high dependency on fossil fuels—by far the largest source of greenhouse gas pollution. They believe that in order to accomplish this, the country must achieve greater energy efficiency, use more renewable energy resources (such as solar, wind, geothermal, and ocean energy), and design cars that run cleaner and get much better gas mileage.

Global warming is already having a significant impact on the world's wildlife. Defenders of Wildlife warns that unless action is taken now to cut the emissions of heat-trapping greenhouse gases, the consequences will be enormous. Their scientists warn that up to 37 percent of the Earth's plants and animals could be extinct by 2050. Human communities and industries will lose the healthy ecosystems that enhance humans' quality of life, produce valuable natural resources, help purify the air and water, and perform other life-sustaining services. They also warn that cutting greenhouse gas emissions alone is not enough. The delayed impact of the gases already in the atmosphere guarantees additional warming and its consequences for decades to come. The effort must be global to reduce the impacts of the global warming already set in motion and make addressing these impacts on ecosystems and wildlife a top global priority.

Global Warming around the World

As new discoveries are made, technology advances, and populations grow and move. The world is increasingly affected by globalization, consumption, and waste production—especially by the developed countries (such as the United States)—and the effects are felt worldwide. Unsustainable practices and uses of natural resources through activities such as deforestation, degradation of water sources, greenhouse gas emissions, mismanagement of rangeland and invasive species, and environmentally unfriendly agricultural practices affect the environment and human health on a global basis. The activities of these developed countries also have negative effects on the developing countries through sea-level rise, drought, flooding, severe weather, and altered rainfall regimes and climate patterns.

Because individuals' decisions on whether or not to be environmentally responsible in the magnitude of the "carbon footprint" they create as it contributes to the global effect, it is important to understand how global warming is affecting different regions around the world—it is important to understand the global scale of this problem. This chapter provides an overview of the effect global warming is having around the world today as well as what is expected in the future.

As reported in the *New York Times* on June 11, 2008, an organized group of industrialized countries (the United States, Britain, Canada, France, Germany, Italy, Japan, and Russia) called the "Academies of the Group of 8 Industrialized Countries," as well as Brazil, China, India, Mexico, and South Africa, issued a statement on June 10, 2008, calling on the industrialized countries of the world to lead a "transition to a low-carbon society" and to aggressively move to limit impacts from changes in climate that are already under way and impossible to stop.

The statement was officially posted by the National Academy of Sciences in the United States, which also urged the "Group of 8" to halve global emissions of greenhouse gases by 2050 and "take maximum efforts" to obtain that goal. Possible solutions to achieving the goal would be as follows:

- adopting new energy technologies;
- changing human behavior and the demand for energy;
- investing in renewable energy research;
- continuing research on carbon sequestration;
- conducting large-scale reforestation;
- exploring "artificial geo-engineering" to stabilize the climate.

Each region worldwide will have to confront the multitude of serious effects from global warming and climate change.

EUROPE, RUSSIA, AND ASIA

According to the World Resources Institute, because of historic land use and settlement, Europe's natural ecosystems are highly disturbed and fragmented. Therefore, they are especially vulnerable to the adverse effects of global warming. During this past century, for example, most of Europe experienced temperature increases that were larger than the global average. In addition, precipitation increased in the north but decreased in the south.

The mountainous regions are providing documentable evidence of global warming. Pronounced evidence is seen in the visible, widespread retreat of mountain glaciers in the Alps. Multiple plant and animal species are migrating northward in latitude and upward in elevation on mountains. Many are also changing the timing of their activities to coincide with the earlier arrival of warm springlike temperatures. One of the major downfalls in this region is that the landscape is so fragmented; the less adaptive species may not be able to fully adapt, putting their existence in peril.

The expansive Asian region includes a wide range of climate regimes. Its polar, temperate, and tropical climates provide a home to more than 3 billion people. Major effects of global warming include the melting of mountain glaciers such as those in the Himalayas (including Mount Everest). Permafrost regions are expected to thaw, such as those in Siberia; and northern forests are expected to shift even farther northward. A major factor in the future of this area is that both China and India are becoming rapidly industrialized. As they continue to modernize and use increasing amounts of fossil fuels, adding to greenhouse gas emissions, if regulations are not put in place to control emissions and pollution, the negative effects will accelerate the damage from global warming, thereby enhancing climate change.

According to a report in *National Geographic News* in June 2002, a team sponsored by the United Nations Environment Programme (UNEP) found signs that the landscape of Mount Everest had changed significantly since the mountain was first successfully climbed in 1953 by Sir Edmund Hillary and Tenzing Norgay. Today, the same glacier that once came close to Hillary and Norgay's first camp has retreated 3 miles (5 km). A series of ponds that used to be near Island Peak has merged into one long lake.

Roger Payne, a sports and development director at the International Mountaineering and Climbing Federation and one of the expedition's leaders, reported: "It is clear that global warming is emerging as one, if not the, biggest threat to mountain areas. The evidence of climate change was all around us, from huge scars gouged in the landscapes by sudden, glacial floods to the lakes swollen by melting glaciers. But it is the observations of some of the people we met, many of whom have lived in the area all their lives, that really hit home."

Tashi Janghu Sherpa of the Nepal Mountain Association said he "had seen quite rapid and significant changes over the past 20 years in the ice fields and that these changes appeared to be accelerating."

In addition, according to the United Nations, dozens of mountain lakes in Nepal and Bhutan are so swollen from melting glaciers that they could burst in the next few years and devastate many Himalayan villages, presenting yet another negative effect of global warming. A team of scientists from the United Nations Environment Program and remote-sensing experts from the International Center for Integrated Mountain Development in Katmandu, Nepal, determined that in the next five years or so, the Himalayas could experience intense flooding as mountain lakes overflow with water from melting glaciers and snowfields. As reported in *National Geographic News* on May 7, 2002, the lives of tens of thousands of people who live high in the mountains and in downstream communities could be at severe risk as the mud walls of the lakes collapse under the pressure of the extra water. Major loss of land and other property would aggravate poverty and hardship in the region.

According to Surendra Shrestha, the Asian regional coordinator of the UNEP's Division of Early Warning and Assessment, "Our findings indicate that 20 glacial lakes in Nepal and 24 in Bhutan have become potentially dangerous as a result of climate change." She also points out that those are the lakes the UNEP studied. The unknown factor could be even worse; there could be lakes not yet identified that are even more dangerous and unstable.

Based on data from the Natural Resources Defense Council, Union of Concerned Scientists, Environmental Defense, National Environmental Trust, World Resources Institute, and World Wildlife Fund, the following list portrays some of the current effects of global warming in Europe, Russia, and Asia:

- *Central England*: Cold days are declining; hot days are increasing. In 1772, there were an average of four days a year of temperatures above 68°F (20°C); in 1995, there were 26 days.

- *Caucasus Mountains, Russia*: Half of all the glacial ice has disappeared in the last 100 years.
- *Austria*: This area has experienced record glacial retreat. The glacial ice is more reduced today than it has been over the past 5,000 years.
- *Spain*: Half of the glaciers in existence in 1980 are now melted.
- *Southeast Europe and the Middle East*: These areas have experienced widespread heat waves. Temperatures climbed as high as 111°F (43.8°C) in Turkey, Greece, Romania, Italy, and Bulgaria. In Bulgaria alone, the 100-year records for daily maximum temperature have been broken at more than 75 percent of the observing stations. In Armenia, 2000 was the hottest summer of the century. Continental Europe warmed 1.4°F (0.8°C) during the past century, with the last decade the hottest on record.
- *Greenland*: There has been rapid thinning of the Greenland ice sheet in coastal areas, especially of the outlet glaciers, documented during the 1990s. The coastal land ice loss is from a combination of increased melting during warmer summers, high snow accumulation rates feeding the outlet glaciers, and increased rates of melting at the bottom of glaciers due to ocean warming.
- *United Kingdom*: Toads, frogs, and newts are spawning nine to 10 days earlier now. From 1971 to 1995, 31 percent of 65 bird species studied showed significant trends toward earlier egg laying, moving up the date by an average of 8.8 days. Oak trees are leafing out earlier. Over a 20-year period, many birds have extended the northern margins of their ranges by an average of about 12 miles (19 km).
- *Austria*: Over a 70- to 90-year period, alpine plants in the Alps moved higher up mountain slopes in response to an increase in average annual temperature.
- *Mediterranean*: This area has experienced intense drought and fires.
- *Europe*: A study of European plants from 1959 to 1993 shows that spring events (such as flowering) have advanced six days and autumn events (such as the changing of leaf colors) have been delayed about five days. This is in response to Europe warming

1.4°F (0.8°C) over the past century. The growing season is also starting about eight days earlier.

- *Southeastern Norway*: The year 2000 was the wettest year on record. Precipitation in northern Europe has increased 10 to 40 percent in the last century.
- *Samos Island, Greece*: Fires caused by dry, droughtlike conditions and record-breaking heat burned one-fifth of the island in July 2000. Temperatures climbed to 104°F (40°C) in some areas.
- *Llasa, Tibet*: June 1998 was the warmest month on record. Temperatures were above 77°F (25°C) for 23 days.
- *Garhwal Himalayas, India*: The Dokriani Barnak Glacier retreated 66 feet (20.1 m) in 1998, even though the region experienced a severe winter. The Gangorti Glacier is retreating 98 feet (30 m) per year. At this rate, scientists predict the loss of all central and eastern Himalayan glaciers by 2035.
- *Southern India*: A massive heat wave occurred in May 2002, when temperatures rose to 120°F (49°C), resulting in the highest one-week death toll on record.
- *Tibet*: Ice core records from the Dasuopu Glacier indicate that the last decade and last 50 years have been the warmest in 1,000 years.
- *Mongolia*: A 1,738-year tree-ring record from remote alpine forests in the Tarvagatay Mountains indicates that 20th-century temperatures in this region are the warmest of the last millennium.
- *Chokoria Sundarbans, Bangladesh*: Rising ocean levels have flooded about 18,550 acres (7,500 hectares) of mangrove forest during the past three decades.
- *Bhutan*: As Himalayan glaciers melt, glacial lakes are swelling and in danger of catastrophic flooding.
- *India*: Glaciers in the Himalayas are retreating at an average rate of 50 feet (15 m) per year. The Khumbu Glacier, a popular climbing route to the summit of Mt. Everest, has retreated more than three miles (5 km) since 1953.
- *Siberia*: Large expanses of tundra permafrost are melting. In some regions, the rate of thawing of the upper ground is nearly 8 inches (20 cm) per year. Thawing permafrost has already damaged more than 300 buildings in the cities of Norilsk and Yakutsk.

EYEWITNESS ACCOUNTS

The effects of global warming are already being felt worldwide. Many peoples' livelihoods are being affected by changes in rainfall patterns and seasonal shifts in temperature, such as earlier spring warm up or longer, drier summers. Those who spend their days outside in the agricultural and horticultural industries, in forestry capacities or vineyards or ranching, have already witnessed firsthand the effects of global warming. Their accounts reflect similar incidences in other parts of the world and are a preview of more to come.

Piacenza, Italy

Giuseppe Miranti from Piacenza, Italy, is the owner of a bioagricultural company called Aziende Agricole Miranti, where he produces fruit and vegetables as well as organic cereal and livestock farming. He is also a beekeeper. He serves as the "Italian Climate Witness" for the World Wildlife Fund. He has already experienced negative impacts from global warming, and he cautions that global warming is a reality—its impacts are visible all over Italy and the rest of Europe.

The cultivation of honey has changed. Due to warmer temperatures, flowers are blooming at unusual times, which makes the bees change their behavior. As a result, the level of activity has slowed down. The biological cycle of bee parasites is another serious problem. The parasites live longer and are more persistent because of the warmer climate. This has a negative impact on the bee population and honey production.

Over time, bees have been able to change to natural changes in the environment, but Giuseppe is now concerned that the bees will not be able to adapt fully to man-made climate change. Because of the concern that bees may not be able to adapt to humans' drastic altering of the natural environment, beekeepers are currently testing new products and methods, adjusting the nutrition in order to ensure that bees are in good condition when the flowers start blooming.

Italy is facing several physical factors as a result of global warming: A heat wave in 2003 caused 20,000 people to die. Almost 2,000 forest fires were reported during the same summer, and drought-related agricultural

(continues)

(continued)

damage cost around 5 billion euros. In 2005, the country experienced another heat wave and severe drought. The Italian government warned that 1 million people were at serious health risk when temperatures reached 40°F (4.4°C). Since 1996, Italy has also become drier. Rainy days have decreased 14 percent, but thunderstorms have actually become more intense.

Bavaria, Germany

Georg Sperber has been a forester for 30 years in the rich, lush forests of Germany. He says, "The forests I have worked in have changed over these years. Especially in the past 20 years. I have seen changes that were remarkable in their nature and intensity. I believe climate change is the main reason. You hear a lot about global warming in the media, but out there in the woods you can feel the difference, even without knowing the alarming facts of climate science. The 1990s have been the warmest decade in climate history, and this was obvious to anyone who lives in touch with nature. In my forests, the consequences for spruce trees are especially dramatic. Spruce is the backbone of the German forest industry, covering 28 percent of Germany's forests. However, higher average temperatures and more frequent droughts due to climate change weaken these trees.

"They are under attack from bark beetle populations, which have massively increased because of the warming. And over past years storms—worse in intensity due to climate change—have wreaked havoc on spruce forests. Rainfall patterns have also changed significantly. Rainfall used to peak in spring and early summer when the plants needed the extra water most. But since the 1990s, this peak has moved to autumn.

"Climate change is the biggest challenge mankind is facing. Currently we are about to put a huge burden on the shoulders of our children and grandchildren. We are absolutely aware that we are doing it, but we know that we shouldn't be doing it."

Source: World Wildlife Fund

AFRICA AND OCEANIA

The geographic position and enormous size of the African continent have endowed it with an abundance of rich, diverse ecosystems ranging from the snow and ice fields of Kilimanjaro to the lush, tropical rain forests near the equator to the dry, imposing Sahara. Africa has the lowest per capita fossil energy use of all the major world regions, relying mainly on primitive fuels such as wood. In contrast, however, Africa is one of the most vulnerable areas on Earth to the negative effects of global warming because its widespread poverty severely limits its ability to adapt to climate change. Because of this, the impacts from global warming that are already under way are being felt, crippling a region unable to fight back. Already the effects of global warming are hitting Africa hard, such as the spreading of disease to populations; the melting of glaciers in the mountains, eliminating a much-needed water source; warming temperatures in drought-prone areas, leading to lack of water resources and contaminated water, and resulting in widespread illness; sea-level rise and the resulting flooding of coastal villages, leaving residents homeless; and the bleaching of coral along the coastlines, destroying fragile marine ecosystem habitats.

As outdoor adventurer Vince Keipper climbed Mount Kilimanjaro in January 2003, he was shocked by what he witnessed on the way up the mountain's Western Breach. "The sound brought our group to a stop. We turned around to see the ice mass collapse with a roar. A section of the glacier crumbled in the middle, and chunks of ice as big as rooms spilled out on the crater floor."

"Just connect the dots," says Ohio State University geologist Lonnie Thompson. "If things remain as they have, in 15 years Kilimanjaro's glaciers will be gone."

Oceania is the world region that encompasses the lush tropical rain forests of Indonesia and the region including New Zealand and the interior deserts of Australia. This area of the world is subject to the effects of El Niño and other ocean phenomena. The small island nations and coastal regions—where most of the population of this region is located—are very vulnerable to the increasing coastal flooding and erosion due to rising sea level as a result of global warming. In addition, warming ocean temperatures in the past few decades have damaged many of the region's fragile coral reefs, putting some of the Earth's most diverse, delicate ecosystems at risk.

According to the World Wildlife Fund, human-induced global warming was a key factor in the severity of the 2002 drought in Australia. Higher temperatures caused a drastic increase in the evaporation rates from soil, rivers, and vegetation. As a result, the higher temperatures and resultant drier conditions created a greater brush fire danger than previous droughts. Drought severity also negatively affected Australia's agricultural production as daytime maximum temperatures from March through November averaged 2.7°F (1.6°C) higher than normal.

Professor David Karoly, a professor of meteorology at Monash University, remarked: "The higher temperatures experienced throughout Australia in 2002 are part of a national warming trend over the past 50 years which cannot be explained by natural climate variability alone." He also points out that the current trend in Australian temperature since 1950 is matching the climate model studies of how temperatures respond to increased greenhouse gases in the atmosphere. He believes that this is the first drought in Australia in which the impact of human-induced global warming can be clearly observed.

Anna Reynolds, World Wildlife Fund's Australia Climate Change Campaign manager, says, "Most of this warming is likely due to the increase in greenhouse gases in the atmosphere from human activity such as burning fossil fuels for electricity and transport and from land clearing."

The Natural Resources Defense Council, Union of Concerned Scientists, Environmental Defense, National Environmental Trust, World Resources Institute, and World Wildlife Fund have identified the following global warming impacts in this region:

- *Kenya*: To date, 92 percent of Mount Kenya's largest glacier, Lewis Glacier, has melted in the past 100 years.
- *Mount Kilimanjaro*: The glacial ice is projected to disappear by 2020. Since 1912, 82 percent has already disappeared, with about one-third of it melting in just the last dozen years.

- *Senegal*: Sea-level rise is causing the loss of coastal land at Rufisque on the south coast of Senegal.
- *Kenya*: Hundreds of people died from a malaria outbreak in the summer of 1997 in the Kenyan highlands, where the population had previously been unexposed.
- *Indian Ocean, Seychelles Islands, and Persian Gulf*: Extensive coral reef bleaching is destroying this fragile ecosystem at an alarming rate.
- *South Africa*: In January 2000, South Africa experienced one of the driest Decembers on record, and temperatures rose more than 104°F (40°C). These conditions fueled extensive fires along the coast in the Western Cape Province.
- *New Zealand*: The warmest February on record occurred in 1998 with temperatures near 67°F (19.4°C). In addition, the average elevation for glaciers has shifted upslope by more than 300 feet (91.4 m) over the past 100 years.
- *American Samoa, Fiji, Papua New Guinea, Philippines, the Indian Ocean, and Australia's Great Barrier Reef*: These areas are victims of extensive coral bleaching due to warming ocean waters.
- *Indonesia*: Wildfires burned 2 million acres (809,371 hectares) of primary forest, including critical areas of the Kalimantan orangutan habitat in 1998.

THE ARCTIC AND ANTARCTICA

According to a report in *National Geographic News* on December 13, 2002, the six largest Eurasian rivers are currently dumping an excessive amount of freshwater into the Arctic Ocean—much more than they were even a few decades ago.

Bruce Peterson at the Marine Biological Laboratory in Woods Hole, Massachusetts, says, "The mechanism is most likely due to increased precipitation as forecast by global climate models."

This finding supports the long-held belief that freshwater runoff into the ocean would increase in the Arctic as a result of global warming. Other factors that may also be contributing to the situation include changes in ice and permafrost melt or changes in the seasonality of precipitation and runoff. The major concern with an increase of freshwater being deposited to the Arctic Ocean is that it could negatively affect ocean circulation patterns in the North Atlantic, thereby altering the heat distribution characteristics of the major currents. Potentially, the influx could slow down or shut off the North Atlantic Deep Water formation (referred to as the Ocean Conveyor Belt), an important current that supplies large amounts of warm water to the North Atlantic region.

Research of the amount of freshwater currently being discharged into the Arctic Ocean is currently about 31 cubic miles (128 km^3) greater than it was in the 1930s. According to Bruce Peterson, "One thing that can slow the circulation of the Ocean Conveyor Belt is if you add freshwater to that area of the North Atlantic. According to several ocean circulation models, if enough freshwater is added to the Arctic Ocean, it will eventually shut down the Conveyor Belt. If you stop the process, you stop the conveyor that brings warm water north. The paradoxical result would be a cooling of northern Europe."

Stefan Rahmstorf of the Potsdam Institute for Climate Impact Research in Germany says, "Most of the rest of the world gets warmer when the conveyor is shutdown. Ocean currents don't generate heat, they just move it around. A shut down would probably enhance global warming, since shutting down the conveyor would also reduce the ocean's uptake of CO_2, so that more of our emissions remain in the air."

Evident impacts to Antarctica from global warming will likely occur first in the northern (warmer) sections of the continent, where the summer temperatures approach the melting point of water: 32°F (0°C). In recent years, some of the ice shelves in the northernmost portions of Antarctica, such as in the Antarctic Peninsula, have been collapsing. This has principally occurred during the gradual warming trend since 1945. One of the largest areas of concern for scientists studying Antarctica is the West Antarctic ice sheet on the main continent. If it were to melt, weaken, and collapse, it could raise sea levels by 19 feet (5.8 m).

According to a report in *National Geographic News* on March 6, 2003, when a huge floating shelf of ice hinged to the northern end of the Antarctic Peninsula disintegrated in January 1995, several glaciers that were backed up into it surged toward the sea. Its discovery provided the first absolute evidence that glacial surge follows an ice shelf collapse.

This is a key finding because it allows scientists to focus on the theory that ice shelves act as "dams" that prevent inland glaciers from slipping into the seas. For more than 30 years, scientists debated whether or not ice shelves acted as barriers for inland ice. Prior to this discovery, many models assumed the type of ground surface it was traveling across dictated the rate of glacial flow. The increased flow after the collapse of the shelf shed new light on understanding the mechanisms of movement and dynamics of the Antarctic glaciers. It is important because if the same disintegration scenario were to happen with major ice shelves such as the Ross Ice Shelf or the Filchner-Ronne, the entire West Antarctic Ice Sheet could collapse and cause global sea levels to rise 16 feet (5 m), assuming ice shelves were the only thing holding the glaciers back from the sea. This debate gives scientists an opportunity to study glacial surge and retreat in Antarctica and how their delicate balance affects ocean levels. It also allows them to get a better idea of the delicate interaction and stability between ice shelves and glacier retreat.

In another study conducted by Isabella Velicogna, a research scientist at Colorado University at Boulder's Cooperative Institute for Research in Environmental Sciences, based on data from two NASA satellites called the Gravity Recovery and Climate Experiment (GRACE), the amount of water pouring annually from the ice sheet into the ocean—equivalent to the amount of water the United States uses in three months—is causing global sea level to rise by 0.016 inch (0.04 cm) a year. Velicogna says, "The ice sheet is losing mass at a significant rate. It's a good indicator of how the climate is changing. It tells us we have to pay attention."

The Natural Resources Defense Council, Union of Concerned Scientists, Environmental Defense, National Environmental Trust, World Resources Institute, and World Wildlife Fund have identified the following global warming impacts in this region:

- *Antarctic Peninsula*: The warming rate is five times the global average. Since 1945, it has experienced a warming of about 4.5°F (2.5°C). The annual melt season has increased by two to three weeks in the past 20 years.
- *Antarctica*: Ice shelf disintegration has occurred on the Larsen A ice shelf, where 770 square miles (1,994 km^2) disintegrated in

January 1995. The Larson B and Wilkins ice shelves collapsed during 1998–99, destroying 1,150 square miles (2,978 km^2) of ice. In February 2002, the northern section of the Larsen B ice shelf, an area of 1,250 square miles (3,250 km^2), disintegrated in a period of 35 days, making it the largest collapse of the past 30 years. The ice thickness has also been decreasing. Permanent ice cover of nine lakes on Signey Island has decreased by 45 percent since the 1950s.

- *Southern Ocean*: There is a strong warming trend in the waters around Antarctica. From the 1950s to the 1980s, the temperatures have risen 0.3°F (0.17°C).

SOUTH, CENTRAL, AND NORTH AMERICA

The ecosystems in South America are vulnerable to changes in water availability brought about by global warming. Higher global temperatures along with more frequent El Niño events bring increased drought. In addition, melting glaciers in the Andes threaten the future water supply of mountain communities. The population of South America is very dependent on the continent's natural resources, making any compromises to them by global warming—such as drying of the rangelands at the foothills of the Andes, damage of the fisheries off the coast of Peru, or destruction of the Amazon rain forest—devastating to human survival and health. Signs of global warming are already occurring at high elevations (in glacial retreat and shifting ranges of disease-carrying mosquitoes) and along the coasts (in rising sea levels and coral bleaching).

The climate of Central America plays a large factor in both the social and economic conditions through its impacts on agriculture, tourism, and human health. The devastating effects of the 1997–98 El Niño serves as an example of what the future will look like if global warming increases. For example, that year saw forest fires burn out of control and high sea surface temperatures bleach the corals in the adjacent seas. If warming continues, future changes in the frequency of extreme events such as hurricanes, droughts, and floods may damage important export crops such as bananas. It may also negatively affect human settlements on unstable hillsides and encourage the outbreak of infectious diseases such as malaria and dengue fever.

The North American continent spans a multitude of climatic zones: from the lush subtropical ecosystems of Florida to the frozen landscapes of Alaska. This area represents a diverse range of ecosystems, many of which are at risk due to global warming; risks include polar warming in Alaska, coral reef bleaching in Florida, animal range shifts in California, glaciers melting in Montana, and marsh loss along the East Coast.

The Natural Resources Defense Council, Union of Concerned Scientists, Environmental Defense, National Environmental Trust, World Resources Institute, and World Wildlife Fund have identified the following global warming impacts in this region:

- *Andes Mountains, Peru*: Glacial retreat is accelerating. The Qori Kalis glacier was retreating 13 feet (4 m) each year between 1963 and 1978. By 1995, however, the rate had increased up to 99 feet (30 m) per year.
- *Tropical Andes*: There has been an increase in the average annual temperature. It has increased by about 0.18°F (0.1°C) per decade since 1939.
- *Argentina and Venezuela*: Glaciers are melting at accelerated rates. Glaciers in Patagonia have receded almost a mile (1.5 km) over the last 13 years. Of six glaciers in the Venezuelan Andes in 1972, only two remain, and scientists predict that these will be gone within the next 10 years.
- *Andes Mountains, Colombia*: Disease-carrying mosquitoes are spreading dengue and yellow fever viruses at higher elevations.
- *Caribbean, Galapagos, and Ecuador*: Massive coral reef bleaching is occurring.
- *Pampas region, Argentina/Uruguay*: The worst flooding on record in 2001 occurred. Nearly 8 million acres (3.2 million hectares) of land in the Pampas region were flooded after three months of high rainfall.
- *Bermuda*: Rising sea level is leading to saltwater inundation of coastal mangrove forests, causing massive die-offs.
- *Andes Mountains*: Disease-carrying mosquitoes are spreading dengue and yellow fever viruses from 3,300 feet (1,006 m) to 7,200 feet (2,195 m).

- *Mexico*: Dengue fever is being spread to higher elevations. It is now appearing at 5,600 feet (1,707 m).
- *Central America*: Dengue fever is being spread to 4,000 feet (1,219 m).
- *Mexico and Nicaragua*: Wildfires have burned massive areas of land.
- *Bering Sea and Arctic Ocean*: Sea ice is shrinking.
- *Canadian Rockies*: Glaciers such as the Athabasca Glacier are retreating.
- *Western Hudson Bay, Canada*: Polar bears are becoming stressed. Decreased weight in adult polar bears and a decline in birthrate since the early 1980s has been attributed to the earlier spring breakup of sea ice.
- *Banks Island, Canada*: Species are expanding their natural ranges. The Inuit report seeing species common much farther south that were never previously seen on the island, such as robins and barn swallows.

GLOBAL WARMING EFFECTS IN THE UNITED STATES

The effects of global warming in the United States include some regional commonalities and also important differences in climate-related issues and consequences. Water is an overriding key issue is all the regions, yet the specific changes and impacts may vary from region to region. Because each major region in the United States will react in a different way, management plans must be tailored to best respond to the specific characteristics of the area in the coming decades and century.

Both short- and long-term decisions about infrastructure, land use, and other planning issues must be made now in order to be prepared for the future. Research is needed in order to be able to make sound decisions of feasibility, effectiveness, and costs of the adaptation strategies.

The Northeast

The Northeast is characterized by diverse waterways, extensive shorelines, and varied landscapes. It includes mountains, forests, and some of the most densely populated areas in the country, such as New York City. Global warming is expected to increase some of the weather

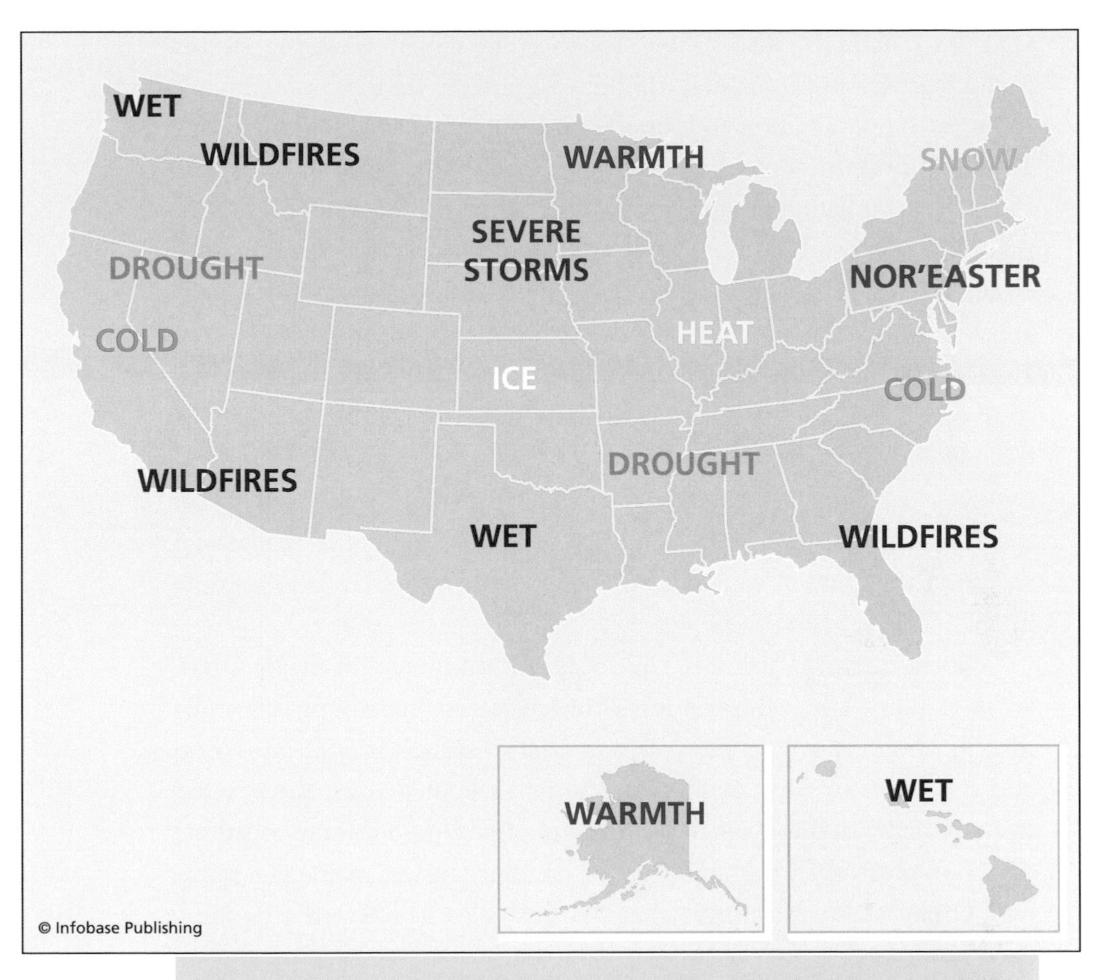

This map represents significant U.S. weather and climate events for 2007. *(NOAA)*

extremes. According to the U.S. Global Change Research Program, the warming projected by climate models over the next several decades suggests increases in rain episodes over frozen ground or rapid snow melting events that can increase flooding. Over the coming century, winter snowfalls and periods of extreme cold will probably decrease. Of major concern for this area are changes in the intensity and frequency of hurricanes.

The estuaries, bays, and wetlands of the Northeast coastal zone are highly valued as unique ecosystems, major recreational sites, migratory waterfowl habitats, and fishery sources. As global warming causes ocean temperatures to rise, it will increase stresses on these ecosystems. Changes in precipitation and runoff will negatively affect coastal salinity. This is important because both temperature and salinity have significant effects on fish populations, as well as human and ecosystem health. As sea levels continue to rise, wetlands and marshes located along the coasts will be destroyed as ocean water invades and contaminates these areas. Because there are already other stresses on the ecosystems in the Northeast, such as the flow of excess nutrients into the bays from industrial, urban, and rural sources, as well as pollution, additional stresses from global warming will exacerbate the already stressed environment. One way to help protect the coastal wetlands from the effects of global warming and sea-level rise is to utilize the lands adjacent to the coastal wetlands as a zone of wetland expansion, enabling wetlands to migrate inland as the sea level rises.

Global warming will also have a significant impact in major urban areas. In many cities, the existing infrastructure (water supply, communication systems, energy delivery, and waste disposal) is aged, worn out, and has not been well maintained. In its weakened state, the stresses from global warming (such as drought, flooding, extreme weather) could be more than the systems can handle, causing major emergencies. Other stresses in urban areas that may also be affected are crime, chronic air quality problems, and inadequate power supply to meet peak energy demands. One positive aspect is that if winters are less severe, there will be fewer winter stresses and casualties. According to the U.S. Global Change Research Program, the major potential consequences of global warming include the impacts of rising sea level and elevated storm surges on transportation systems, increased heat-related illness and death associated with temperature extremes, increased ground-level ozone pollution associated with warming, and the impact of precipitation and evaporation changes on relatively inflexible water supply systems. Their suggestions for solutions include changing water supply management programs, replacing aging infrastructure with more climate-resilient systems, strengthening water quality and air quality controls to minimize the compounding of climate impacts, and using early warning systems and measures such as changing roofing colors and adding shade trees to limit urban heat that can contribute to heat-related stresses and deaths.

There also will be limitations in the recreation industry, principally the ski industry, activities at beaches and freshwater reservoirs, and sightseeing in the forested areas, as well as hiking, camping, and fishing. Suggested adaptations strategies will need to reflect a regional shift in recreational activity as people make trade-offs in terms of the type, location, and season of their activities.

In addition, there will be changes in species composition, possibly changes in predator-prey relationships, changes in pest types and populations, invasive species, and in key species that are truly characteristic of a region or are of economic significance. The lobster population is expected to migrate farther northward in order to stay in the cool water environment that it thrives in. Warmer waters are expected to start limiting the trout population, especially in locations such as Pennsylvania. Changes in species mix and the introduction of climate-driven invasive species are also likely to induce unanticipated feedbacks on ecosystems.

The Southeast

The Southeast—the Sunbelt—is a rapidly growing area of the United States. A great majority of this growth has occurred on the coastal areas of the region. It is expected to grow another 40 percent between now and 2025.

The Southeast is a very productive region of the country, supplying one-fourth of the United States's natural crops. Often referred to as the "wood basket," the region produces half of the country's timber supplies. It also produces a large portion of the nation's fish, poultry, tobacco, oil, coal, and natural gas. About half of the remaining wetlands in the lower 48 states are found in the Southeast; and more than 75 percent of the nation's annual wetland losses over the past 50 years occurred in this region. The region has a wide range of ecosystems, and species diversity is high.

Climate in the region varies on a decadal scale. The period from the 1920s to the 1940s was warm but was followed by a cool trend in the 1960s. Since the 1970s, temperatures have been increasing, with the

1990s temperatures as warm as the peaks in the 1920s and '30s. Annual rainfall has increased 30 percent or more over the past 100 years across Mississippi, Arkansas, South Carolina, Tennessee, Alabama, and parts of Louisiana. There have also been strong El Niño and La Niña effects in the region.

Climate model projections for this region using the Hadley model (a climate model developed by the United Kingdom's Hadley Centre for Climate Prediction and Research) scenario simulates less warming and a significant increase in precipitation of about 20 percent. Some of the climate models suggest that the rainfall associated with El Niño and the intensity of droughts during La Niña phases will be intensified as atmospheric CO_2 increases.

There are several weather-related stresses on human populations in this region. Over the past 20 years, more than half of the nation's costliest weather-related disasters (such as hurricanes) have occurred in the Southeast. Across the region, intense precipitation has increased over the past 100 years, and this trend is projected to continue. There are also concerns about heat waves and drought. In 1998, there was a heat wave and drought that resulted in at least 200 deaths and caused damages in excess of $6 billion. Air quality is another major issue. As temperatures rise and increased emissions enter the atmosphere from power generation facilities, the ground-level ozone is increased, further degrading air quality. Increased flooding in low-lying coastal counties from the Carolinas to Texas is also likely to adversely affect human health—floods are the leading cause of death from natural disasters in the region and nationwide.

In the past, traditional ways to adapt included building structures such as flood levees, elevated structures (such as homes on stilts), and modifying building codes to account for flooding. Unfortunately, with global warming, this approach no longer provides adequate protection, especially in the coastal regions where sea-level rise continues to increase the likelihood of storm-surge flooding in virtually all southeastern coastal areas. It is crucial that these areas develop strategies for hazard preparedness in order to protect themselves from this very real threat.

Regarding agricultural crop yields, the lower Mississippi Valley and Gulf Coast areas are expected to have decreased yields, but the northern Atlantic Coastal Plain is expected to experience increased yields.

Neighborhoods in New Orleans, Louisiana, on September 2, 2005, remain flooded as a result of Hurricane Katrina. Rescue crews go into neighborhoods, looking for residents unable to get out of their houses. *(Jocelyn Augustino.FEMA)*

In order to deal with possible negative impacts, producers have the option to switch crops or vary planting dates, patterns of water usage, crop rotations, and the amounts, timing, and application methods for fertilizers and pesticides. Models indicate that farmers, except those in the southern Mississippi Delta and Gulf Coast areas, will probably be able to mitigate most of the negative effects and possibly benefit from changes in CO_2 and moisture, which enhance crop growth. Agriculture may also be helped through the development of improved varieties of crops designed to withstand changed weather conditions.

Forested areas of both pines and hardwoods are expected to increase productivity, mostly in the northern regions. Canadian models predict that savannas and grasslands due to decreased soil moisture and wildfire will replace part of these forests. Future management plans could include shifting timber harvesting northward in order to take stress off the forests in the southern reaches of the region. Other adaptation strategies include the use of more drought-hardy strains of pine and genetic improvements that could increase water use efficiency or water availability. Research and the creation of up-to-date forest management regimes are going to be important as the climate changes. As the climate gets hotter and drier, an aggressive fire management strategy will become critical.

Maintaining water quality is another important issue. Currently, water quality is challenged by intensive agricultural practices, urban development, coastal processes, and mining activities. As global warming progresses, these stresses on water quality will become even more pronounced. Higher temperatures will reduce the dissolved oxygen levels in the water. As flooding becomes more of a problem, floodwaters fouled with sewage, rotting farm animal carcasses, fuel, and chemicals swamping water treatment plants and contaminating public water supplies will become more common.

Storm surge damage in coastal areas due to sea-level rise will also become more of a problem. Projected impacts include: loss of barrier islands and wetlands that protect coastal communities from storm surges, reduced fisheries productivity as coastal marshes and submerged grassbeds are displaced or eliminated, and saltwater intrusion into surface and groundwater supplies. If there are changes in precipitation, it will alter the freshwater inflow into estuaries and alter the salinity patterns that determine the survival and distribution of coastal plant and animal communities.

The Midwest

The Midwest is characterized by farming, manufacturing, and forestry. The Great Lakes form the world's largest freshwater lake system, providing a major recreation area as well as a regional water transportation system with access to the Atlantic Ocean via the St. Lawrence Seaway. This region encompasses the headwaters and upper basin of the Mississippi River and most of the length of the Ohio River, both critical water sources and means of industrial transportation providing an outlet to the Gulf of Mexico. This area contains some of the richest farmland in the world and produces most of the nation's corn and soybeans. It also has significant metropolitan centers such as Chicago and Detroit. The North Woods are a large source of forestry products and have the advantage of being situated near the Great Lakes, which provide convenient transportation.

During the 20th century, the northern portion of the area warmed by almost 4°F (2°C), while the southern portion, along the Ohio River valley, has cooled by about 1°F (0.5°C). Annual precipitation has increased, in some places up to 20 percent, causing a higher number of rainy days.

Climate models predict that temperatures will increase throughout the Midwest during the 21st century at a greater rate than has been observed in the 20th century. Even over the northern portion of the region, where warming has been the greatest, an accelerated warming trend is projected, with temperatures increasing by 5–10°F (3–6°C). Models also predict that there will be a substantial increase in evaporation, causing a soil moisture deficit, a reduction in lake and river levels, and more droughtlike conditions in much of the region. Unusually heavy and extreme precipitation events are also expected.

Water resources also will be affected. Water levels, supply, quality, and water-based transportation and recreation are all sensitive to climate change. Even though there is a projected increase in precipitation, increased evaporation due to higher summer air temperatures will most likely cause reduced levels in the Great Lakes. The U.S. Global Change Research Program ran 12 different models, and 11 of them projected a one- to five-foot (0.3 -1.5 m) decrease in lake levels for the Midwest. A five-foot (1.5 m) reduction would cause up to a 40 percent reduction in outflow to the St. Lawrence Seaway. Lower lake levels also cause reduced hydropower generation downstream—a clean, renewable energy source—with reductions of up to 15 percent by 2050. As water availability declines, there is also the concern of increased national and international tension related to the increased pressure for water use from the Great Lakes. Decisions will have to be made between the United States and Canada as to how to divide the water-use rights.

Higher temperatures will be a critical factor in the highly populated urban areas because of the urban heat island effect. As an example of this, in 1995 a heat wave hit Chicago, causing more than 700 heat-related deaths. In addition, during heat waves, air pollutants become trapped near the Earth's surface, causing unhealthy air quality. With the likelihood of an increase in extreme precipitation, there also will be an increased risk of water-borne diseases and insect- or tick-borne diseases such as West Nile virus or mosquito-borne St. Louis encephalitis.

Recreational activities will shift as cold-season recreation such as skiing, ice skating, snowmobiling, and ice-fishing are reduced and warm-season recreation such as swimming, hiking, and golf are expanded. Higher temperatures, however, will probably impair summer recreational activities, making people more susceptible to heat stroke.

Another key issue for the Midwest is agriculture. This region is a key producer of crops not only for the United States but also for the world. In the past, it has been able to adapt to changes in climate, and scientists at the U.S. Global Change Research Program think that it will continue to adapt in the future. As temperatures warm, the growing season will lengthen, possibly allowing the opportunity to "double crop," the practice of planting a second crop after the first is harvested. The CO_2 fertilization effect will enhance plant growth and most likely contribute to higher crop yields in the region. The northern portions are expected to have the largest increases. Additional productivity could be met if plant-breeding programs create new varieties more tolerant to new growing conditions. Years with droughts and flooding will still cause significant decreases in production, however.

There also will be significant changes in the natural ecosystems. Because of higher temperatures and increased evaporation, forested areas will become more susceptible to pests and diseases. Temperate forests are expected to migrate northward into the present-day boreal forest habitat. There will be additional forest growth on the current forest margins due to the increase in CO_2. Most of the climate models predict that higher air temperatures will cause greater evaporation, which will reduce soil moisture, making conditions ripe for wildfires. Drier soils will also put greater stress on both deciduous and coniferous trees, making them susceptible to disease and pest infestation, increasing tree mortality rates.

As the water temperatures in lakes increase, changes in freshwater ecosystems will occur. Current coldwater fish species such as trout will not be able to survive. Warmer water species such as bass and catfish will inhabit and thrive in the ecosystem instead. As heavy precipitation events increase, there will be increased runoff of excess nutrients such as nitrogen and phosphorus from fertilizer into the lakes and rivers. This will cause the growth of algae, depleting the water of oxygen, negatively effecting the fish and other life that inhabits the water. Lower lake levels will also have a negative impact on adjoining wetlands.

The Great Plains

Farming and ranching are the primary land uses of the Great Plains, but urban areas provide housing and jobs for about two-thirds of the region's people. Native ecosystems and agricultural fields exist along with small rural communities and expanding metropolitan centers. This area of the country produces much of the nation's grain, meat, and fiber, including more than 60 percent of the wheat, 87 percent of the sorghum, and 36 percent of the cotton. The region also carries more than 60 percent of the nation's livestock, including both grazing and grain-fed-cattle operations.

Temperatures in the northern and central Great Plains have risen more than 2°F (1°C) in the last 100 years, with increases up to 5.5°F (3°C) in parts of Montana, North Dakota, and South Dakota. Over the same time period, annual precipitation has decreased by 10 percent in eastern Montana, North Dakota, eastern Wyoming, and Colorado. In the eastern portion of the Great Plains, precipitation has increased by more than 10 percent. Texas has experienced significantly higher intensity rainfall. The snow season ends earlier in the spring, reflecting the greater seasonal warming in winter and spring.

Climate models project that temperatures will continue to rise throughout the region, with the largest increases in the western parts of the Great Plains, causing significant heat stress for both people and livestock. On a seasonal basis, more warming is projected for the winter and spring seasons than for the summer and fall. Precipitation is expected to increase in the north, but decrease in the west adjacent to the Rocky Mountains. According to the U.S. Global Change Research Program, a

25 percent decrease is projected in the Oklahoma Panhandle, northern Texas, eastern Colorado, and western Kansas. Even though there will be an increase in precipitation for parts of the Great Plains, increased evaporation is also projected due to rising air temperatures, which will result in a net soil moisture decline for large parts of the region, negatively affecting agricultural activities.

Water supply, demand, allocation, storage, and quality are all affected by global warming issues. Farming and ranching currently use more than half of the region's water resources. Groundwater pumping for irrigation has currently depleted aquifers in parts of the Great Plains by withdrawing water much faster than it can currently be recharged. Under today's irrigation demands, water table levels are dropping in southern portions of the region. The projected global warming-induced changes in water resources are expected to exacerbate the current competition for water among the agricultural sector; natural ecosystems; and urban, industrial, and recreational users. Possible solutions include switching to crops that use less water, retiring marginal lands from farming, and adopting conservation tillage methods.

Invasive species are currently a serious challenge in both the native ecosystems and agricultural systems. Global warming is projected to alter the current biodiversity of the region, and the possible migration of invasive species across the Great Plains is a significant concern.

The West

A variable climate, diverse topography and ecosystems, increasing human population, and a rapidly growing and changing economy characterize the West. Scenic landscapes range from the coastal vistas of California to the intimidating deserts of the Southwest to the alpine meadows of the Rocky and Sierra Nevada Mountains. Since 1950, the West has quadrupled its population, expanding urban areas such as Los Angeles, Las Vegas, Salt Lake City, Santa Fe, Albuquerque, Tucson, and Phoenix. Numerous national parks and monuments exist in the West—such as Zion National Park, Arches National Park, Death Valley, Canyonlands, and the Grand Staircase-Escalante National Monument—and attract millions of tourists from around the world. Because of the extreme population growth and development, this region faces multiple stresses such as air quality problems, urbanization, and wildfires. The most significant problem, however, is the availability of water resources. This region includes the characteristically dry Southwest. Water is usually consumed far from where it originates. Competition for water among agricultural, urban, power consumption, recreational, environmental, and other uses are intense, with water supplies already not ample enough in many areas.

Another factor in the West is its variable climate. Historically, it has experienced exceptionally wet and dry periods. During the 20th century, temperatures in the West have risen 2–5°F (1–3°C). The region has also had increases in precipitation, with increases in some areas greater than 50 percent. Some areas, however, such as Arizona, have become drier and experienced more droughts. The length of the snow season has decreased by 16 days from 1951 to the present in California and Nevada. Extreme precipitation events have increased, causing flooding and landslides, such as those in California.

Modeling efforts conducted by the U.S. Global Change Research Program project annual average temperature increases 3–4+°F (2°C) by the 2030s and 8–11°F (4.5–6°C) by the 2090s. The models also project increased precipitation during winter, especially over California, where runoff is projected to double by the 2090s. In these climate scenarios, some areas of the Rocky Mountains are projected to get drier. The models project more extreme wet and dry years.

The future of the West's water resources are of critical concern because they are so sensitive to climate change. The semiarid West is dependent upon a vast system of engineered water storage and transport, such as along the Colorado River, and is governed by complex water rights laws. Much of the water supply comes from snowmelt. The problem with global warming is that snowpacks are expected to decrease, and earlier spring warming will alter the amount and timing of peak flows, upsetting the balance and management of water resources. In some areas, resource managers from the Bureau of Reclamation believe it is likely that current reservoir systems will be inadequate to control earlier spring runoff and maintain supplies for the summer. The situation will become even more critical as populations continue to grow.

In order to adapt to these changes brought on by global warming, improved technology, planting of less water-demanding crops, and other conservation measures will become increasingly important. Transferring water across basins and water users and the integration of surface and groundwater use could also serve as an adaptation strategy for resource managers facing water shortages.

Climate models predict an increase in plant growth, a reduction in desert areas, and a shift toward more woodlands and forests in many parts of the West. They also project an increase in wildfires and a rise in air pollution. Eventually, as CO_2 concentrations level off, the fertilization effect would decrease, leading to a decline in forest productivity. A drier climate would also reduce forest productivity.

Ecosystems will also be affected. The diverse topography coupled with landscape fragmentation and other development pressures in the West will probably make it difficult for many species to adapt to climate change by migrating. The U.S. Global Change Research Program believes it is likely that ecosystems such as alpine ecosystems will disappear entirely from some places in the region. They also believe that some species may be able to successfully migrate up mountain slopes to higher elevations. One prominent factor resource managers must consider, however, is that non-native invasive species have already stressed many Western ecosystems and are likely to make adaptation to climate change much more difficult for native species. Global warming is also likely to increase fire frequency. As long as year-to-year variation in precipitation remains high, fire risk is likely to increase, whether the region gets wetter or drier because fuel loads tend to increase in wet years as a result of increased plant productivity and are consumed by fire in dry years.

Tourism, which is a prominent, growing component of the West's economy, is strongly oriented to the outdoors and is extremely sensitive to climate. Higher temperatures will prolong summer season activities such as backpacking but make the winter season activities such as skiing shorter. Changes in the distribution and abundance of vegetation, fish, and wildlife will also affect recreational activities. These effects will cause the tourism industry to become more diversified and force it to provide activities commensurate with the current climate.

The Pacific Northwest

The Pacific Northwest encompasses extensive forests, topography that creates abrupt changes in climate and ecosystems over short distances, with mountain and marine environments in close proximity. The Northwest provides about one-fourth of the nation's softwood lumber and plywood. The fertile lowlands of eastern Washington produce 60 percent of the nation's apples and large fractions of its other tree fruit. Population and economic growth of this area has been twice the national rate since 1970, its population nearly doubling during this period. The area provides moderate climate, a high quality of life, and outdoor recreational opportunities, which are becoming increasingly stressed because of the rapid development occurring in the area. Stresses are occurring today from dam operations, land-use conversion from natural ecosystems to metropolitan areas, intensively managed forests, agriculture, and grazing. The result has been in the loss of old-growth forests, wetlands, and native grasslands; increased urban air pollution; extreme reduction of salmon runs; and increasing numbers of threatened and endangered species.

Over the 20th century, the region has grown warmer and wetter. Annual average temperature has increased 1–3°F (0.5–1.5°C) over most of the region. Annual precipitation has also increased across the region by 10 percent on average, with increases reaching 30 to 40 percent in eastern Washington and northern Idaho. The region's climate also shows recurrent patterns of year-to-year variability. The warm years tend to be relatively dry with low stream flow and light snowpack, while cool ones tend to be relatively wet with high stream flow and heavy snowpack. This variability has distinct effects on the ecosystems. Warmer, drier years have summer water shortages, less abundant salmon, and increased probability of forest fires.

The U.S. Global Change Research Program projects in their models that regional warming in the 21st century will be much greater than observed during the 20th century, with an average warming over the region of about 3°F (1.5°C) by the 2050s. By the 2090s, average summer temperatures are projected to rise by 7–8°F (4–4.5°C), and winter temperatures are expected to rise 8–11°F (4.5–6°C). Precipitation is expected to develop a pattern of wetter winters and drier summers, with an overall decrease in water supply. By the 2090s, however, the projected annual average precipitation is expected to increase, ranging from a few percentage points to 20 percent.

Warmer, wetter winters will increase flooding in the rain-fed rivers. Projected year-round warming and drier summers will probably increase summer water shortages in both rain-fed and snow-fed rivers, including the Columbia, because there would be less snowpack and because it would melt earlier. In the Columbia River, there are already serious water allocation conflicts, and the waterway is already vulnerable to water shortages.

In this region, evergreen coniferous forests dominate the landscape. Approximately 80 percent of the area west of the Cascades is covered with coniferous forests. These forests are sensitive to climate variation because warm, dry summers stress them directly by limiting seedling establishment and summer photosynthesis, as well as indirectly by creating conditions favorable to pests and wildfire. The extent, species mix, and productivity are expected to change under the influence of global warming. The amount of change will depend on the interaction of the amount of precipitation, the seasonal water-storage capacity of forest soils, and changes in the trees' water-use efficiency under elevated CO_2. These factors will make proper land management and conservation practices important in order to maintain healthy landscapes and protect natural resources. Adaptation measures can include planting species adapted to the projected climate rather than the present climate; managing forest density to reduce susceptibility to drought stress and fire risk; and using prescribed burning to reduce the risk of large, high-intensity fires. Other practices such as reduced tree cutting, reduced road construction, and the establishment of large buffers around streams are some other ways to promote diversity of plant and animal species in the forest ecosystem.

Sea-level rise is also a concern in the Pacific Northwest. Areas such as Puget Sound, where coastal subsidence is occurring, are especially at risk. These areas are also at risk for winter landslides and increased erosion on sandy stretches of the Pacific Coast. Severe storm surges and erosion are presently associated with El Niño events, which raise sea level for several months and change the direction of prevailing winds.

Global warming is projected to bring similar shifts. Projected heavier winter rainfall is likely to increase soil saturation, landsliding, and winter flooding. All of these changes will likely increase the damage to property and infrastructure on bluffs and beachfronts, as well as beside rivers and waterways.

Alaska

Alaska includes a wide range of physical, climatic, and ecological diversity in its rain forests; mountain glaciers; boreal spruce forest; and expansive *tundra,* peat lands, and meadows. It contains 75 percent by area of the U.S. national parks and 90 percent of wildlife refuges, 63 percent of wetlands, and more glaciers and active volcanoes than all other states combined. There is not much pressure put on the land by humans, but the pressure put on its marine environment from large commercial fisheries is substantial.

Alaska has warmed substantially over the 20th century, especially over the past few decades. Average warming since the 1950s has been 4°F (2°C). The greatest area of warming—approximately 7°F (4°C)—has occurred in the interior of the state during the winter. The growing season has lengthened by more than 14 days since the 1950s. Precipitation has increased about 30 percent since 1968. The observed warming is part of a larger trend through most of the Arctic, which has been corroborated by many independent observations and measurements of sea ice, glaciers, permafrost, vegetation, and snow cover. The most severe environmental stresses in Alaska at present are climate related.

Two models run by the U.S. Global Change Research Program (the Hadley and Canadian models) indicate that as global warming progresses, temperatures will warm 1.5–5°F (1–3°C) by 2030, and 5–12°F (3–6.5°C) (Hadley model) or 7–18°F (4–10°C) (Canadian model) by 2100. The greatest amount of warming is expected to occur in the north during the winter. The models also project precipitation increases in most of the state, reaching 25 percent in the north and northwest, with areas of up to 10 percent decrease along the southern coast. The area is also expected to experience increased evaporation from the warming, which will offset the effects from the increase in precipitation, making the soils drier throughout most of the state.

The key issues stemming from global warming in Alaska will be the thawing of permafrost and the melting of sea ice. Permafrost underlies most of Alaska, and the recent several decades of warming have been accompanied by extensive thawing, causing increased erosion, landslides, sinking of the ground surface, as well as disruption and damage to forests, buildings, and infrastructure. Thawing is projected to accelerate under future warming, with as much as the top 30–35 feet (10 m) of discontinuous permafrost thawing by 2100.

Sea ice off the Alaskan coast is retreating and thinning, with widespread effects on marine ecosystems, coastal climate, and human settlements. Arctic sea ice has decreased 14 percent since 1978, the majority of it melted since the 1990s. Since the 1960s, sea ice over large areas of the Arctic basin has thinned by 3–6 feet (1–2 m), losing about 40 percent of its total thickness. All climate models project large continued loss of sea ice, with year-round ice disappearing completely (in the Canadian model) by 2100. The retreat of sea ice allows larger storm surges to develop, increasing the risk of inundation and increasing erosion on coasts that are also made vulnerable by permafrost thawing. In some regions, shorelines have retreated more than 1,500 feet (400 m) due to erosion over the past few decades. Several Alaskan coastal villages will soon have to be reinforced or relocated before they are destroyed. Loss of sea ice also causes large-scale changes in marine ecosystems, threatening populations of marine mammals and polar bears that depend on the ice for hunting and other activities.

The Gulf of Alaska and Bering Sea support diverse marine ecosystems and represent the nation's largest commercial fishery. Their productivity fluctuates with year-to-year and decade-to-decade climate variability. As global warming progresses, the variability is expected to continue. As fish populations decline, it will have a negative effect on the other animals that feed off them in the food chain.

There will also be increased stress on the subsistence livelihoods in the area. Many native people live in isolated rural communities where subsistence is practiced to gather food. In fact, Alaska's 117,000 rural residents collect about 43 million pounds (20 million kg) of wild food annually, equivalent to 375 pounds (170 kg) per person each year. Fish play a large role in their subsistence diet. Present climate change already poses serious harm to subsistence livelihoods. Reduced snow cover, a shorter river ice season, and thawing of permafrost all obstruct travel to harvest wild food. The retreat and thinning of sea ice, with associated stress on marine mammal and polar bear populations and increased open-water roughness, have made hunting more difficult, more dangerous, and less productive. In the long term, projected ecosystem shifts are likely to displace or change the resources available for subsistence, requiring communities to change their practices or move. Shifts in the composition of tundra vegetation may decrease nutrition available for caribou and reindeer. In addition, the invasion of the tundra by boreal or mixed forest will probably limit the range of caribou and musk ox.

As this chapter has pointed out, global warming is already having an effect on daily life worldwide, but when temperatures rise a few degrees, the current inconveniences could give way to dangerous conditions and even death. There is no safe haven from global warming—it is a global phenomenon. Given what scientists already know about global warming and its effects on the environment, it is not to early to begin taking mitigation and adaptation measures to deal with the challenges right now. It is critical to continue climate-impact research at local, regional, national, and global scales and consider the full range of stresses that can affect an ecosystem, such as climate change and variability, land use change, air and water pollution, and many other human and natural impacts. In order to achieve the most comprehensive results, scientists worldwide representing multiple scientific disciplines—such as those in the IPCC—need to continue to work together to solve the problem of global warming.

Mitigation and Adaptation to Climate Change

As the Earth's atmosphere continues to warm and more people become aware and educated about global warming—its effects and ramifications for the future—efforts worldwide are being made to reduce its impacts, to find ways to mitigate the situation, and to adapt to the present environment as well as prepare for the future. This chapter focuses on the impacts of global warming, human mitigation, and adaptation to the effects of global warming, focusing on ways to help solve the problem, to reduce what impacts are possible, to find ways to be environmentally responsible, and to learn to cope in a positive way to permanent change.

IMPACTS OF GLOBAL WARMING

According to a report in the *New York Times* in January 2008, an article appearing in the journal *Nature Geoscience* reported that warming water temperatures and wind patterns consistent with global warming have been discovered to be eroding the ice sheets of Antarctica faster than anyone had previously thought. Up until recently, Antarctica has been one of the "inconsistencies" that opponents of global warming have used to make a case against it—saying that because notable melting is not occurring there, global warming is not a serious issue. This is no longer the case. Although the loss is not massive at this point, the study does suggest that if the trend continues, it could cause global sea levels to rise much more rapidly than originally predicted by climate-change scientists. To the proponents of global warming, this adds an increased urgency to cut back the emissions of CO_2 and other greenhouse gases—many of which have an even greater global warming potential (GWP).

In particular, West Antarctica's huge ice shelf (which is roughly the size of Texas) is losing ice 60 percent faster today than it was just 10 years ago. Based on this information, the IPCC warns that if nothing is done to decrease greenhouse gas emissions, the world's oceans could rise more than 2 feet (0.6 m) by 2100.

Another impact of global warming that has not had much exposure in past news, but is quickly coming to the forefront, is a disaster facing the polar regions. According to a report in *LiveScience,* a phenomenon that could "damage structures in northern areas, reconfigure towering mountains, and alter biology" has become a pressing issue.

As northern, high-latitude winters continue to become increasingly milder, dangerous changes are occurring. Although not visible at first, the process can occur over a period of time before its effects manifest themselves: the melting of the permafrost layer. Nearly one-fourth of the land in the Northern Hemisphere has a layer of permafrost beneath it—a layer of perennially (occurring all year) frozen ground.

Global warming is changing this balance, however. The areas that are frozen seasonally have decreased by 15 to 20 percent during the past century. According to Tingjun Zhang of the University of Colorado at Boulder, "In the last 20 years, the decrease is more dramatic. The change is real. It's happening."

Regions in Russia have been documented where 16 inches (40 cm) below the surface have warmed 1°F (0.6 °C) in average soil temperature. In the extreme northern areas of permafrost where it is melting, it is realigning roads, railroad tracks, toppling houses, cracking founda-

The Trans-Alaska pipeline extends 800 miles (1,287.5 km)—posts are designed to keep permafrost frozen so as to support the pipeline. Topped with fanlike aluminum radiators, the posts absorb cold from the winter air. When the air temperature is, for example, 40 below, the posts take on the same temperature, making the soil around them 40 below too, which keeps the permafrost from melting. The pipeline was built in a zigzag pattern to allow the pipe to expand and contract. *(Patrick J. Endres, AlaskaPhotoGraphics.com)*

tions, causing landslides and mudflows, breaking pipelines, and damaging other infrastructure.

The effects are not just confined to the polar regions, either. In the United States, 80 percent of its soil freezes every year. Changes to this cycle will have detrimental effects such as damage to agricultural lands, native plants, and carbon exchange mechanisms between vegetation and the atmosphere.

Frederick Nelson, a geographer at the University of Delaware, says, "There is widespread evidence that global warming is responsible for the observed changes in seasonally frozen soil and permafrost. Thawing permafrost will profoundly affect biological activity in ways that are not fully known."

When the freezing period does not last as long, making the soil thaw earlier because of global warming, deeper zones in the permafrost region of the soil end up thawing as well. Each year of mild winters, the zone of thawing reaches deeper and deeper, destroying the permafrost. This makes the ground swell and expand when it freezes and contract and shrink when it melts, causing uneven movements on the ground's surface. Over time, this expanding/contracting cycle causes structural damage in urban areas to railroads, roads, and buildings. In natural areas, it can cause landslides and mudslides.

This situation has been addressed by the Qinghai-Xizang railroad in Tibet. The railroad is 695 miles (1,118 km) long, and half of it is built above 13,000 feet (3,962 m). Half of its length is located on zones of permafrost, and according to Tingjun Zhang, this will melt in the coming years because of global warming. Because of this, the railroad stands on a unique insulation system—it rests on a thick layer of crushed rock that covers the permafrost zone.

Charles Harris at Cardiff University in the United Kingdom has documented rockslides at high elevations in the Swiss Alps, which were related to thawing permafrost zones. In fact, in 2008—the warmest summer recorded in the Alps—the slushy "active" layer of permafrost shifted from its long-term 15 feet (4.5 m) to 29 feet (9 m) below the surface of the ground.

According to Harris, "There is likely to be an increase in rock falls and landslides at high altitude sites."

The latest report issued by the IPCC in 2007 outlines the impacts that will require mitigation and adaptation as follows:
Global effects as of 2007:

- enlargement and increased numbers of glacial lakes;
- increasing ground instability in permafrost regions;
- increasing rock avalanches in mountain regions;
- changes in Arctic and Antarctic ecosystems, particularly with polar bears;
- increased runoff and earlier spring peak discharge in many glacier- and snow-fed rivers;

- warming of many lakes and rivers;
- earlier timing of spring events such as leaf unfolding, bird migration, and egg-laying;
- poleward and upward shifts in ranges in plant and animal species;
- shifts and changes in algal, plankton, and fish abundance in high-latitude oceans;
- increases in algal and zooplankton abundance in high-latitude and high-altitude lakes;
- range changes and earlier migrations of fish in rivers;
- increase in ocean acidity.

Regional climate changes as of 2007:

- effects on agricultural and forestry management at Northern Hemisphere higher latitudes, such as earlier spring planting of crops and changes in forest ecosystems due to fires and pests (insects);
- human health issues such as heat-related mortality, infectious diseases, and allergenic pollen;
- disturbance of human activities in the Arctic (transportation, hunting, travel over snow and ice);
- drought in areas such as the Sahelian region of Africa;
- flooding along coastal wetlands and shorelines.

Fortunately, much work has been done worldwide and sufficient progress has been made in understanding the types of impacts humans are now, and will be, facing due to the effects of global warming. Having a solid understanding of the impacts allows action to be taken in the form of mitigation.

MITIGATION STRATEGIES—CARBON CAPTURE AND STORAGE

Mitigation strategies relating to global warming involve taking positive action to reduce greenhouse gas emissions and also to enhance sinks aimed at reducing the extent of global warming. Mitigation is any measure that will reduce greenhouse gas emissions, such as using wind-generated energy instead of a coal-fired plant. Mitigation is taking action to reduce or eliminate the problem before it becomes a problem, not just merely allowing the problem to happen and then adjusting to it.

According to the IPCC, in order to stabilize greenhouse gas concentrations, the following greenhouse gases would have to be reduced by at least the following percent:

Greenhouse Gas Reduction Necessary to Stabilize the Atmosphere

GREENHOUSE GAS	REDUCTION REQUIRED TO STABILIZE ATMOSPHERIC CONCENTRATION
Carbon Dioxide	60%
Methane	15–20%
Nitrous Oxide	70–80%
Source: IPCC	

In a new study conducted by Scott Doney of the Woods Hole Oceanographic Institution (WHOI), a newly developed computer model indicates that the land and oceans will absorb less carbon in the future if current trends of emissions continue, which could mean significant shifts in the climate system.

According to Scott Doney, "Time is of the essence in dealing with greenhouse gas emissions. We can start now or we can wait 50 years, but in 50 years we will be committed to significant rapid climate change, having missed our best opportunity for remediation." He also stressed that the Earth's ability to store carbon in its natural reservoirs is inversely related to the rate at which carbon is added to the atmosphere. In other words, as soon as humans cut greenhouse gas emissions, the easier it will be for the Earth to naturally store carbon. He stresses that the study suggests that land and oceans can absorb carbon at a certain rate, but at some point they may not be able to keep up.

Presently, about one-third of all human-generated carbon emissions have dissolved in the ocean. How fast the ocean can remove CO_2 from the atmosphere depends on atmospheric CO_2 levels, ocean circulation, and mixing rates. The more CO_2 in the atmosphere, the more there can be in the ocean; faster circulation increases the volume of water exposed to higher CO_2 levels in the air, which increases uptake by the ocean.

Global warming, however, will cause ocean temperatures to rise; and warmer water holds less dissolved gas, which means the oceans will not be able to store as much anthropogenic CO_2 as global warming continues. A negative side effect on the oceans of global warming is that increasing amounts of CO_2 in the oceans will increase its acid content. When CO_2 gas dissolves in ocean water, it combines with water molecules (H_2O) and forms carbonic acid (H_2CO_3). The acid releases hydrogen ions into the water. The more hydrogen ions in a solution, the more acidic it becomes. According to the WHOI, hydrogen ions in ocean surface waters are now 25 percent higher than in the preindustrial era, with an additional 75 percent increase projected by 2100.

Carbon Capture and Storage

According to the IPCC, CO_2 capture and storage (CCS) is a process that involves capturing the CO_2 arising from the combustion of fossil fuels (such as in power generation or refining fossil fuels), transport to a storage location, and long-term isolation from the atmosphere.

Before CO_2 gas can be sequestered from power plants and other point sources, it must be captured as a relatively pure gas. According to the U.S. Department of Energy, on a mass basis, CO_2 is the 19th-largest commodity chemical in the United States, and it is routinely separated and captured as a by-product from industrial processes such as synthetic ammonia production and limestone calcinations.

CSS has the potential to reduce overall mitigation costs, but its widespread application would depend on overall costs, the ability to successfully transfer the technology to developing countries, regulatory issues, environmental issues, and public perception. The capture of CO_2 would need to be applied to large point sources such as energy facilities or major CO_2-emitting industries to make it cost effective. Potential storage areas for the CO_2 would be in geological formations (such as oil and gas fields, unminable coal seams, and deep saline formations), in the ocean (direct release into the ocean water column or onto the deep seafloor), and industrial fixation of CO_2 into inorganic carbonates. Storage in geological formations, however, may face risks due to leakage from structural characteristics in the storage location, as discussed previously in chapter 2.

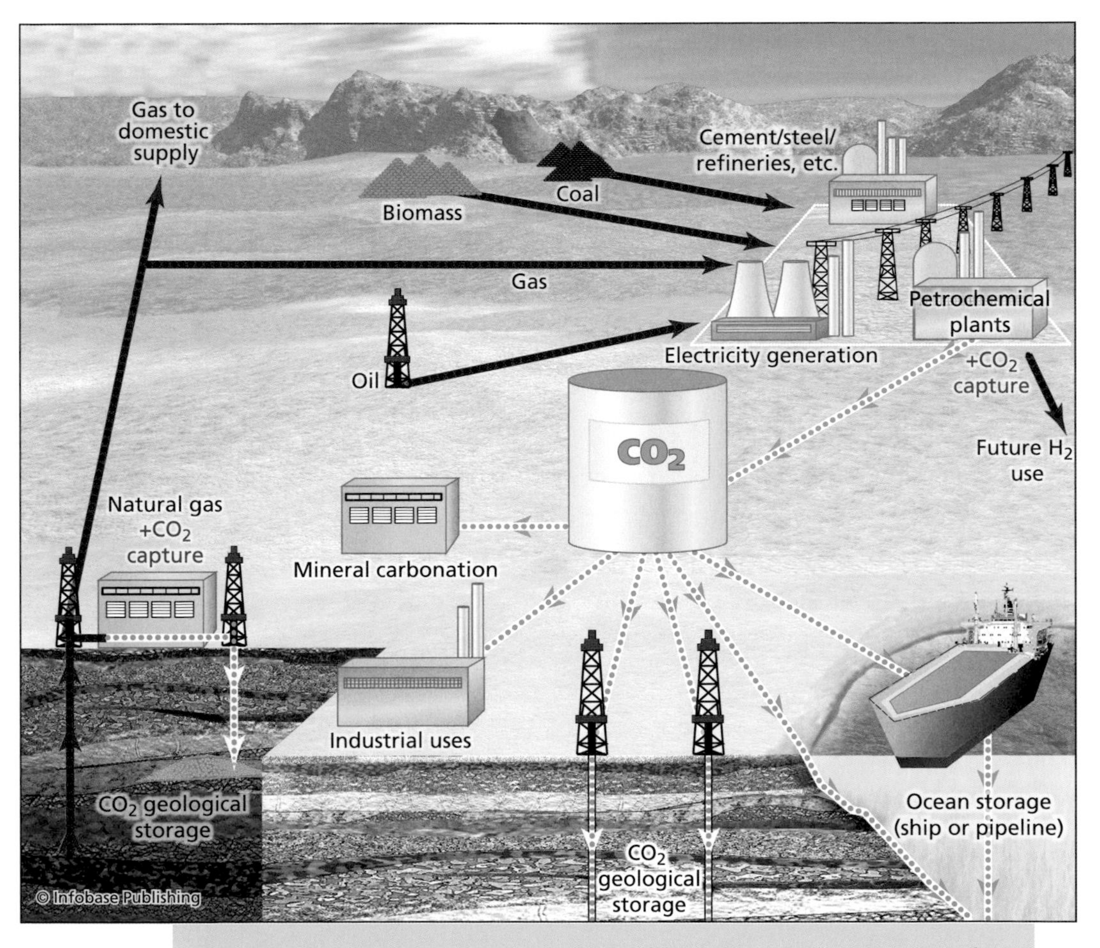

The IPCC's schematic diagram of possible CCS systems, showing the sources for which CCS might be relevant, transport of CO_2, and storage options *(IPCC)*

Current technology captures roughly 85–95 percent of the CO_2 processed in a capture plant. A power plant that has a CCS system (with an access to a geological or ocean storage) uses approximately 10–40 percent more energy than a plant of equivalent output without CCS (the extra energy is for the capture and compression of CO_2). The final result with a CCS is that there is a reduction of CO_2 emissions to the atmosphere by 80–90 percent compared to a plant without CCS.

When CO_2 is captured, it must be separated from a gas stream. Techniques to do this have existed for 60 years. Used in the production of town gas by scrubbing the gas stream with a chemical solvent, CO_2 removal is already used in the production of hydrogen from fossil fuels. This practice helps remove CO_2 from contributing to global warming.

When the CO_2 is transported to its storage site, it is compressed in order to reduce its volume; when it is compressed, it occupies only 0.2 percent of its normal volume. Each year, several million tons of CO_2 are transported by pipeline, by ship, and by road tanker.

Today there are several options for storing CO_2. Initially, it was proposed to inject CO_2 into the ocean, where it would be carried down into deep water and would stay for hundreds of years. In order for any CCS scheme to be effective, however, it needs to sequester huge amounts of CO_2—comparable to what is currently being submitted into the atmosphere—in the range of gigatons per year. Due to the size requirement, the most feasible storage sites are the Earth's natural reservoirs such as certain geological formations or deep ocean areas.

The technology of injecting CO_2 underground is very similar to what the oil and gas industry uses for the exploration and production of hydrocarbons. It is also similar to the underground injection of waste practiced in the United States. In the same manner, wells would be drilled into geological formations, and the CO_2 would be injected. This is also the same method used today for enhanced oil recovery. In some areas, it has been proposed to pump CO_2 into the ground for sequestration while simultaneously recovering oil deposits. There are arguments both for and against this strategy; on one hand, recovering oil would offset the cost of sequestration. On the other hand, burning the recovered oil as a fossil fuel adds additional

CO_2 to the atmosphere, which offsets some of the positive effects of the sequestration.

Two other strategies involve injecting CO_2 into saline formations or into unminable coal seams. The world's first CO_2 storage facility, located in a saline formation deep beneath the North Sea, began operation in 1996.

Other alternatives have been proposed as well, such as using CO_2 to make chemicals or other products, fixing it in mineral carbonates for storage in a solid form such as solid CO_2 (dry ice), CO_2 hydrate, or solid carbon. According to the IPCC, another suggested option is to capture CO_2 from flue gas, using micro-algae to make a product that can be turned into a biofuel.

In order to decide where to find feasible sites for carbon sequestration, it is important to know where large carbon sources are geographically distributed so as to assess their potential. This enables managers to estimate the costs of transporting CO_2 to storage sites.

According to the IPCC, more than 60 percent of global CO_2 emissions originate from the power and industry sectors. Geographically, 66 percent of these areas occur in three principal regions worldwide: Asia (30 percent), North America (24 percent), and Canada/Western Europe (12 percent). In the future, however, the geographical distribution of emission sources is expected to change. Based on data from the IPCC, by 2050 the bulk of emission sources will be from the developing regions such as China, South Asia, and Latin America. The power generation, transport, and industry sectors are still expected to be the leading contributors of CO_2.

Global storage options are focusing primarily on geological or deep ocean sequestration. It is expected that CO_2 will be injected and trapped within geological formations at subsurface depths greater than 2,625 feet (800 m), where the CO_2 will be supercritical and in a dense liquidlike form in a geological reservoir, or injected into deep ocean water so it disperses quickly, or deposited it at great depths on the floor of the ocean with the goal of forming CO_2 lakes. Current estimates place both types of sequestration as having ample potential storage space—estimates range from hundreds to tens of thousands of gigatons (Gt) of CO_2.

Industry contributes to a large portion of greenhouse gases in the atmosphere. It is these types of industry where carbon sequestration options are attractive. *(Nature's Images)*

Geological Formation Sequestration

Many of the technologies required for large-scale geological storage of CO_2 already exist. Because of extensive oil industry experience, the technologies for drilling, injection, stimulations, and completions for CO_2 injection wells exist and are being patterned after current CO_2 projects. In fact, the design of a CO_2 injection well is very similar to that of a gas injection well in an oil field or natural gas storage project. Capture and storage of CO_2 in geological formations provides a way to eliminate the emission of CO_2 into the atmosphere by capturing it from large stationary sources, transporting it (usually by pipeline), and injecting it into suitable deep rock formations. Geologic storage of CO_2 has been a natural process within the Earth's upper crust for millions of years; there are vast reservoirs of carbon held today in coal, oil, gas, organic-rich shale, and carbonate rocks. In fact, over eons CO_2 has been derived from biological activity, igneous activity, and chemical reactions that have occurred between rocks, and fluids and gases have naturally accumulated in the subsurface layers.

The first time CO_2 was purposely injected into a subsurface geological formation was in Texas in the early 1970s as part of an "enhanced oil recovery" effort. Based on the success of this effort, applying the same technology to store anthropogenic CO_2 as a greenhouse gas mitigation option was also proposed around the same time, but not much was done to pursue any actual sequestration. Not until nearly 20 years later, in the early 1990s, research groups began to take the idea more seriously. In 1996, Statoil and its partners at the Sleipner Gas Field in the North Sea began the world's first large-scale storage project. Following their lead, by the end of the 1990s, several research programs had been launched in the United States, Europe, Canada, Japan, and Australia. Oil, coal mining, and electricity-generating companies spurred much of the interest in this technology as a mitigation option for waste by-products in their respective industries.

Since this initial push, environmental scientists—many connected with the IPCC—have become involved in geologic sequestration as a viable option to combat global warming. The significant issues now are whether the technique is (1) safe, (2) environmentally sustainable, (3) cost effective, and (4) capable of being broadly applied.

Geologic storage is feasible in several types of sedimentary basins such as oil fields, depleted gas fields, deep coal seams, and saline formations. Formations can also be located both on and off-shore. Offshore sites are accessed through pipelines from the shore or from offshore platforms. The continental shelf and some adjacent deep-marine sedimentary basins are also potential sites, but the abyssal deep ocean-floor areas are not feasible because they are often too thin or impermeable. Caverns and basalt are other possible geological storage areas.

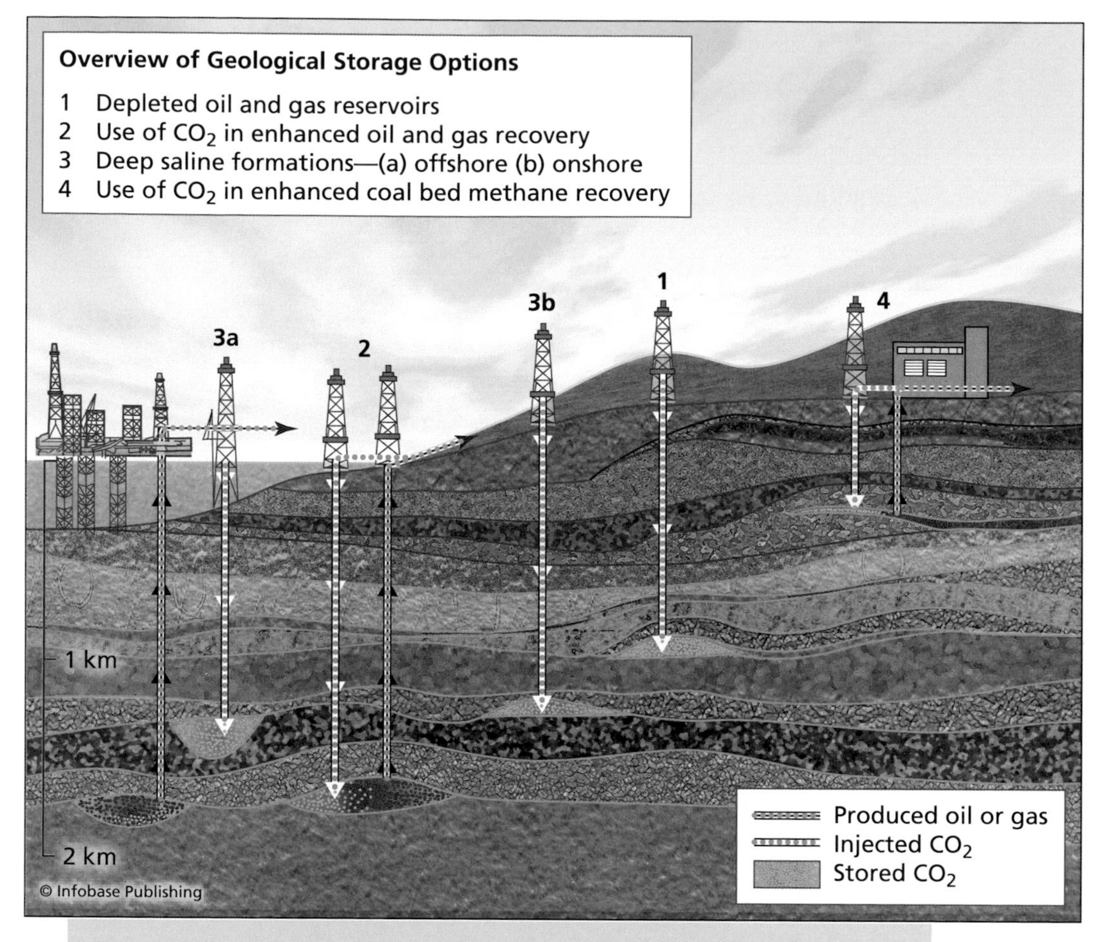

CO_2 can be sequestered in deep underground geological formations. *(IPCC)*

Not all sedimentary basins make good candidates, however. Some are too shallow, some not permeable enough, and others do not have the ability to keep the CO_2 properly contained. Suitable geologic formations require a thick accumulation of sediments, permeable rock formations saturated with saline water, extensive covers of low-porosity rocks to act as a seal, and structural simplicity. In addition, a feasible storage location must also be economically feasible, have enough storage capacity, and be technically feasible, safe, environmentally and socially sustainable, and acceptable to the community. Most of the world's populations are concentrated in regions that are underlain by sedimentary basins.

The following table lists some of the current geological storage projects around the world:

Current Carbon Sequestration Projects within Geological Formations			
COUNTRY	PROJECT NAME	YEAR BEGUN	TOTAL STORAGE
Norway	Sleipner	1996	20 Mt
Canada	Weyburn	2000	20 Mt
Algeria	InSalah	2004	17 Mt
United States	Salt Creek	2004	27 Mt
Japan	Minami-Nagoaka	2002	10,000 t
China	Qinshui Basin	2003	150 t

The most effective geologic storage sites are those where the CO_2 is immobile because it is trapped permanently under a thick, low-permeability seal, is converted to solid minerals, or is absorbed on the surface of coal micropores or through a combination of physical or chemical trapping mechanisms. If done properly, CO_2 can remain trapped for millions of years.

When converting possible local and regional environmental hazards, the biggest danger is that if CO_2 were to seep from storage, human exposure to elevated amounts of CO_2 could cause respiratory problems. This is why these storage facilities are closely monitored.

Announced in November 2007 in the *Salt Lake Tribune*, the U.S. Department of Energy (DOE) is planning on spending $67 million over the next decade on a technological solution for climate change that will be tested in Utah. CO_2 will be pumped into a mile-deep (1.6 km) rock formation under Carbon County for long-term underground storage.

According to Dianne Nielson, Utah governor Jon Huntsman Jr.'s energy adviser, "I don't think such technological methods are as far out as some people say. In Utah, we have a leg up on research and development."

Energy and policy makers in Utah see this opportunity as a way to offset fossil-fuel pollution produced by power plants and energy development.

Brian McPherson, an engineering professor employed by Utah Science Technology and Research (USTAR) to develop commercial applications for new technologies from Utah's major state universities, is currently involved in cutting-edge research involving carbon sequestration engineering.

A total of $88 million is being spent on research and testing primarily in the Farnham Dome formation. This is the location of abandoned oil and gas fields in the southeastern corner of Utah near the Navajo Indian Reservation. The project could expand and cover areas that run from New Mexico to Montana, however, because there are additional underground geological basins that have an enormous storage potential—possibly enough to hold up to 100 years' worth of carbon emissions from major sources.

In this project, liquefied CO_2 will be injected into muddy layers of rock about 5,000 feet (1,524 m) below the desert surface. A layer of impervious shale caps the formation. Farnham Dome already contains a natural pool of CO_2 that formed between 10 and 50 million years ago that was harvested commercially until 1979 for the manufacture of dry ice and soft drinks.

Governor Jon Huntsman is actively looking for ways to reduce Utah's impacts on global warming. He was one who joined California governor Arnold Schwarzenegger in signing the Western Climate Initiative, with a goal to reduce greenhouse gas emissions 15 percent region-wide by 2020.

According to Brian McPherson, "Carbon sequestration will be a good short-term solution for dealing with climate change as the world shifts to renewable energy sources that are now in their infancy."

This project is also a good illustration of teamwork by professionals from various sectors: private, academic, and government. Currently involved are experts from the University of Utah, Utah Geological Survey, Questar Gas, Rocky Mountain Power, Savoy Energy, Blue Source,

Pure Energy Corp., the Navajo Nation, and the New Mexico Institute of Mining and Technology.

Ocean Sequestration

According to the IPCC, various technologies have been identified to enable and increase ocean CO_2 storage. One suggested option is to store a relatively pure stream of CO_2 that has been captured and compressed. The CO_2 could be loaded onto a ship and injected directly into the ocean or deposited on the seafloor. CO_2 loaded on ships could be either dispersed from a towed pipe or transported to fixed platforms feeding a CO_2 lake on the seafloor. The CO_2 must be deeper than 1.9 miles (3 km) because at this depth CO_2 is denser than seawater.

Relative to CO_2 accumulation in the atmosphere, direct injection of CO_2 into the ocean could reduce maximum amounts and rates of atmospheric CO_2 increase over the next several centuries. Once released, it is expected that the CO_2 would dissolve into the surrounding seawater, disperse, and become part of the ocean carbon cycle.

C. Marchetti was the first scientist to propose injecting liquefied CO_2 into waters flowing over the Mediterranean sill into the middepth North Atlantic, where the CO_2 would be isolated from the atmosphere for centuries—a concept that relies on the slow exchange of deep ocean waters with the surface to isolate CO_2 from the atmosphere. Marchetti's objective was to transfer CO_2 to deep waters because the degree of isolation from the atmosphere increases with depth in the ocean. Injecting the CO_2 below the thermocline would enable the most efficient storage. In the short term, fixed or towed pipes are the most viable methods for oceanic CO_2 release because the technology is already available and proven.

One proposed option is to send the CO_2 down as "dry ice torpedoes." In this option, CO_2 could be released from a ship as dry ice at the ocean's surface. If CO_2 has been formed into solid blocks with a density of $1.5tm^{-3}$, they would sink quickly to the seafloor and could potentially penetrate into the seafloor sediment.

Another method, called "direct flue-gas injection" involves taking a power plant fire gas and pumping it directly into the deep ocean without any separation of CO_2 from the flue gas. Costs for this are still prohibitive, however.

It will be possible to monitor distributions of injected CO_2 using a combination of shipboard measurement and modeling approaches. Current analytical monitoring techniques for measuring total CO_2 in the ocean are accurate to about ± 0.05 percent. According to the IPCC, measurable changes could be seen with the addition of 99 tons (90 metric tons) of CO_2 per 0.2 mi^3 (1 km^3). This means that 1.1 gigaton (1 metric gigaton) of CO_2 could be detected even if it were dispersed over an area 2.4 million mi^3 (10^7 km^3 or 5,000 km × 2,000 km × 1 km), if the dissolved inorganic carbon concentrations in the region were mapped out with high-density surveys before the injection began.

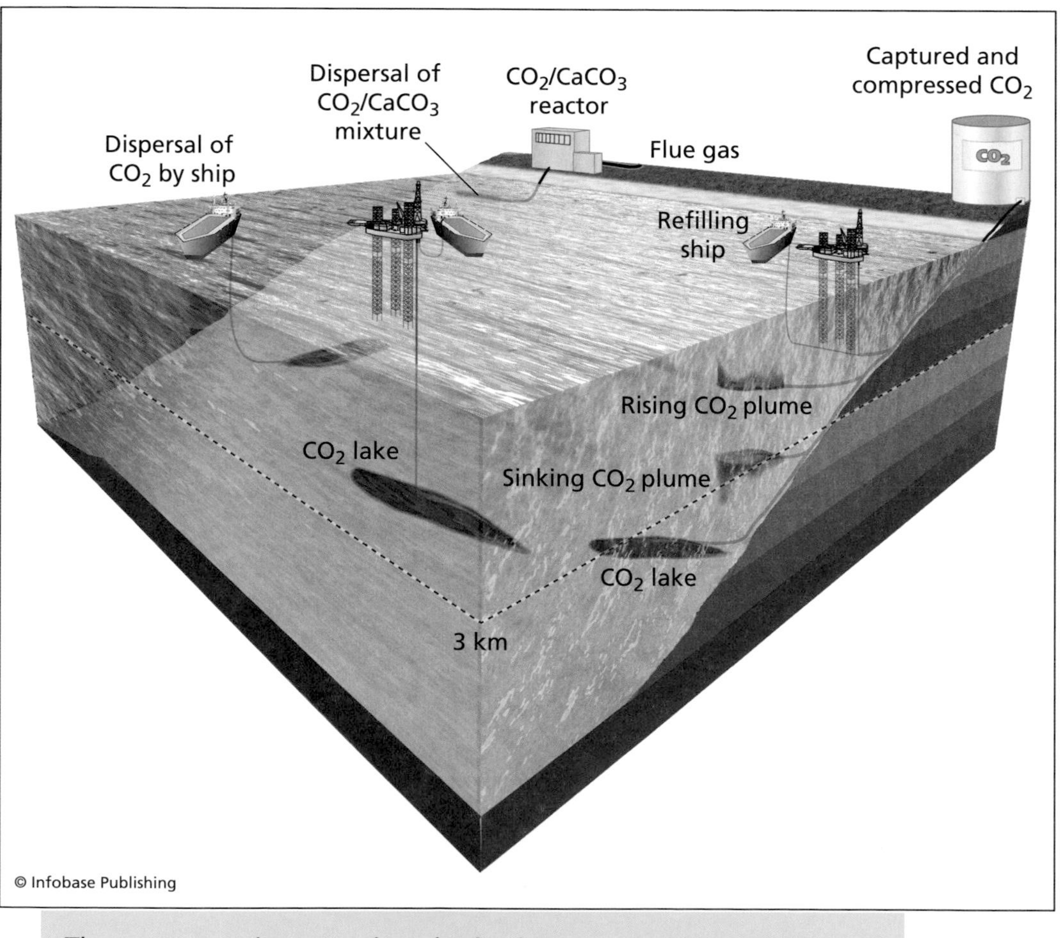

There are several proposed methods of CO_2 sequestration in the world's oceans. *(IPCC)*

In the case of monitoring the injection of CO_2 into the deep ocean via a pipeline, several monitoring techniques are employed. At the point of entry from the pipeline into the ocean, an inflow plume is created of high CO_2/low pH water extending from the end of the pipeline. The first monitoring array consists of sets of chemical, biological, and current sensors; and underwater cameras in order to view the end of the pipeline. An array of moored sensors monitors the direction and magnitude of the resulting plume around the pipe. Monitors are also set along the pipeline to monitor leaks. A shore-based facility provides power to the sensors and is able to receive real-time data. In addition, a forward system monitors the area and can provide data over broad areas very quickly. Moored systems monitor the CO_2 influx, send the information to surface buoys, and make daily transmissions back to the monitoring facility via satellite.

As briefly touched on in chapter 2, a major project to pump CO_2 underground in an ocean sequestration storage project is proposed for the ocean region just off the coast of the U.S. Pacific Northwest. In a report from the *Seattle Times* in July 2008, David Goldberg, a geophysicist at Columbia University's Lamont-Doherty Earth Observatory said, "It's hard to deny the size of the prize."

The research team involved believes it is possible to devise a system where CO_2 emissions from power plants could be captured, liquefied, and pumped into porous basalt layers roughly 0.5 mile (0.8 km) beneath the ocean floor. They currently estimate there is enough storage space in the geologic formation to easily store more than a century's worth of CO_2 emissions from the United States.

Taro Takahashi, also a scientist at Lamont-Doherty on the project, says, "In principle, the type of reservoir we propose on the ocean floor is one of the safest—if not the safest—way of storing liquefied CO_2 for a long, long, time."

The proposal has met with some opposition, however. Some complain that the project will be too costly. Others raise questions about whether there will be possible ecological impacts or seismic hazards. The Juan de Fuca Plate, where the project would be located, is the tectonic plate that subducts under the North American continent and there have been questions raised as to whether the project could trigger earthquakes.

In response, the scientists on the project—David Goldberg and Taro Takahashi—say they have mapped out a 30,000-square-mile area (78,000 km²) that avoids the seismically active regions of the plate.

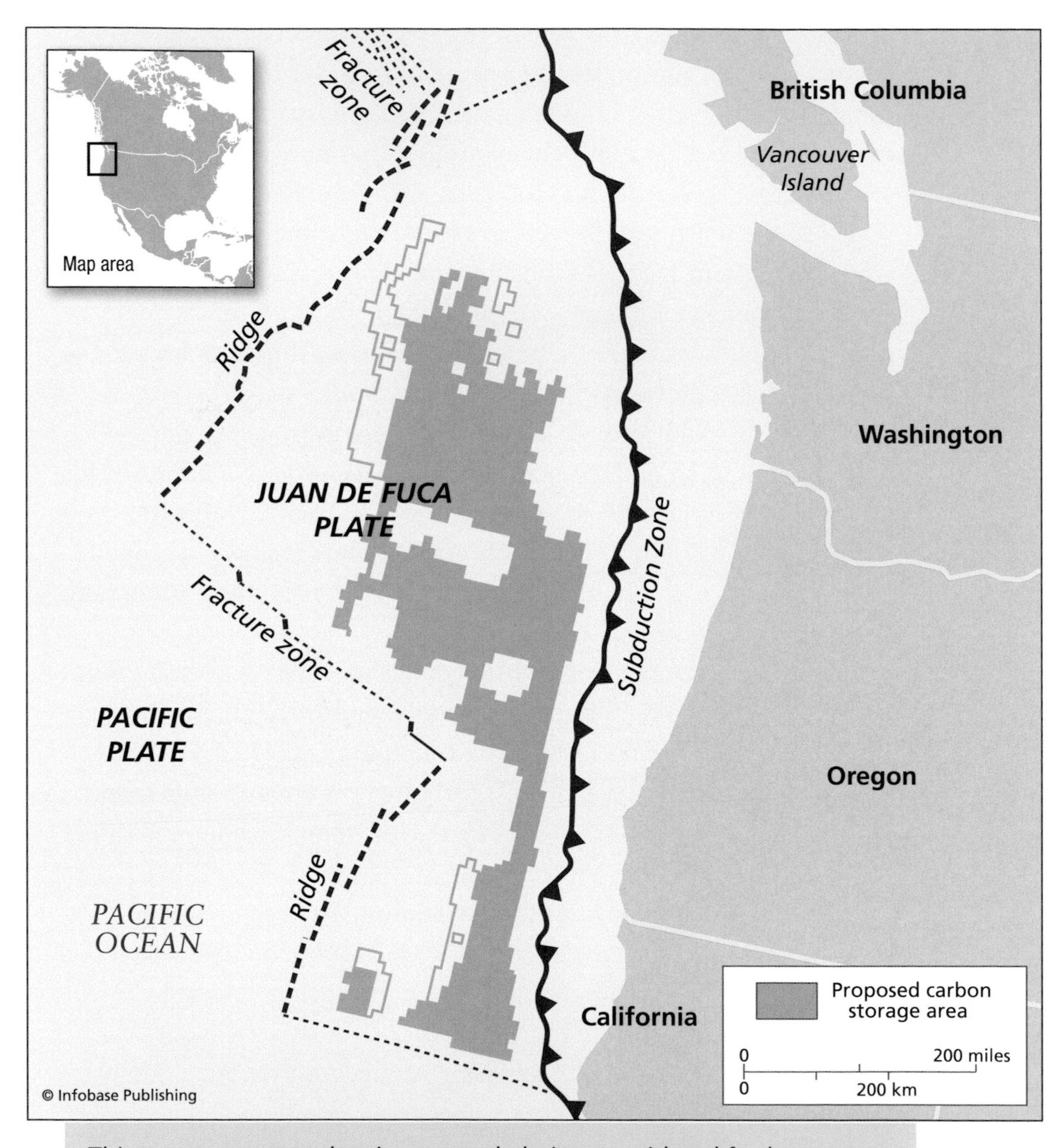

This area represents the site currently being considered for long-term carbon storage off the coast of the Pacific Northwest of the United States. The storage site is located well within the plate boundary to avoid instability.

Other critics are environmentalists who are critical of carbon sequestration in general, claiming that it is just a temporary measure delaying the immediate use of greener energy alternatives, which they feel is a better direction to take.

Angela Slagle, a marine geologist on the project concerned about wildlife habitat, said they would also avoid drilling new hydrothermal vents, where a myriad of unique sea life exists. Instead, the wells would be located more than 100 miles (161 km) from the coastline.

The research team also stresses that undersea basalts can trap and hold CO_2 in several different ways, which in turn provide multiple layers of protection against leaks.

Kurt Zenz House, a Harvard researcher who was one of the first to propose undersea carbon storage, says, "Under immense pressure and cold temperature below the seafloor, CO_2 forms a very dense liquid that is much heavier than seawater. In addition, gravity would prevent the liquefied gas from seeping upward, just as it prevents water in a well from flying into the air."

According to a report in *MSNBC News* in September 2006, Dr. James E. Hansen, a global warming expert at NASA/GISS, says the world has a 10-year window of opportunity to take positive action on global warming to avoid an impending catastrophe. He stresses that it is critical that governments worldwide adopt "an alternative scenario to keep CO_2 emission growth in check and limit the increase in global temperatures to 1.8°F (1°C). I think we have a very brief window of opportunity to deal with climate change . . . no longer than a decade at the most."

In attendance at the Climate Change Research Conference, he stressed that if people continue to ignore the building evidence that global warming is happening and continue their lifestyles in a "business as usual" manner, atmospheric temperatures will rise 3.6–7.2°F (2–3°C) and "we will be producing a different planet." He also warns that under that warmer world drastic changes would be inevitable, such as the rapid melting of ice sheets (which would put most of Manhattan under water), prolonged droughts, severe heat waves, powerful hurricanes in new areas, and the "likely extinction" of 50 percent of the species on Earth.

MITIGATION—ONE STEP AT A TIME: HEADING IN THE RIGHT DIRECTION

According to an article in the *New York Times* in November 2007, the United States could shave up to 28 percent off the greenhouse gases it emits for a relatively inexpensive cost and a few small technological innovations. Most of the reductions could come about through steps that would also lower energy bills for both individuals and businesses and should be done just on the notion that they make good sense, according to energy experts at McKinsey and Company. Reported as well in a June 17, 2008, article on *NewGeography.com,* McKinsey and company developed a "blueprint" that shows how the United States could reduce those emissions by 2030 while "maintaining comparable levels of consumer utility." According to McKinsey, this means "no change in thermostat settings or appliance use, no downsizing of vehicles, home or commercial space and traveling the same mileage." They stress that to be effective and to avoid reducing the standard of living, efforts to reduce GHG emissions must be based upon sound economic analysis. The starting point is an evaluation of strategies to determine the least expensive in terms of cost per ton removed and the least invasive to the way people live and work.

Jack Stephenson, the director of the study, says most of the changes are simple: changes in lighting, and the heating and cooling of buildings would have a significant impact. As an extra incentive, they would also save energy payers money.

A major step consumers can make when purchasing items such as furnaces, stoves, ovens, washers, dryers, televisions, and computers is to

ADAPTATION STRATEGIES

In a report issued by NASA, due to the increasing challenges caused by climate change and global warming, several scientists and policy makers in the United States have come together to take a part in the newly established United States National Assessment on the Potential Consequences of Climate Variability and Change—called the National Assessment for short.

buy energy-efficient products. Even if they cost a little more initially, the savings involved usually pay for themselves via lower energy bills, and the environment benefits as well.

Ken Ostrowski, leader of the project report team, says, "What the report calls out is the fact that the potential is so substantial for energy efficiency. Not that we will do it, but the potential is just staggering here in the U.S. There is a lot of inertia, and a lot of barriers. The country can do the job with tested approaches and high-potential emerging technologies, but doing the work will require strong, coordinated, economy-wide action that begins in the near future."

The report also suggested solutions such as rewriting regulations for the utilities, enabling them to make as much promoting conservation as in selling energy. It also supports the notion of a "broad public education program focusing on wasteful energy consumption."

Taking the lead in one aspect is Delta Airlines. Beginning in spring 2007, the airline gave customers the opportunity to become involved in reforestation efforts by making donations to their Conservation Fund when they booked their airline tickets. The following December, Continental Airlines created a "carbon offsetting calculator" where travelers are able to view the "carbon footprint" of their flight and make a donation to a nonprofit group called Sustainable Travel International. While there, they can choose one of four "offsetting portfolios" that range from reforestation efforts to renewable energy projects. Other airlines such as Air Canada and British Airline are starting to follow and offer similar services.

The National Assessment currently consists of 16 separate regional projects. Project leaders are charged with assessing their region's most vulnerable aspects—the resources that would be affected most by global warming. These include resources such as water supply and quality, agricultural productivity, and human health issues. Once potential impacts are identified, strategies are proposed and developed to cope and adapt to global warming impacts should they occur.

Michael MacCracken, head of the coordinating national office, says, "The goal of the assessment is to provide the information for communities as well as activities to prepare and adapt to the changes in climate that are starting to emerge."

The more successful that mitigation strategies are toward the effects of global warming today, the less human populations will have to adapt in the short and long term. According to the Pew Center on Global Climate Change, however, in recognition that the climate system has a great deal of inertia and is increasing, mitigation efforts alone are now insufficient to protect the Earth from some degree of climate change. Even if extreme measures to combat global warming were taken immediately to slow or even stop emissions, the momentum of the Earth's climate is such that additional warming is inevitable. Some of the warming that is unstoppable now is due to emissions of greenhouse gases that were released into the atmosphere decades ago. Because of this, humans have no choice but to adapt to the damage that has already been done.

According to the Pew Center, adaptation is not a simple, straightforward issue for humans or ecosystems. Each system has its own "adaptive capacity." In systems that are well managed (in developed countries such as the United States), wealth, the availability of technology, responsible decision-making capabilities, human resources, and advanced communication technology help tremendously in successful adaptation to climate change. Societies that are able to anticipate environmental changes and plan accordingly ahead of time are also more likely to succeed.

The ability of natural ecosystems to successfully adapt is another issue, however. While biological systems are usually able to adapt to environmental changes and inherent genetic changes, the timescales are usually much longer than a few decades or centuries (such as the case with global warming). With changes in climate, even minor changes can be detrimental to natural ecosystems. An example of this is the polar bear in the Arctic. Today sea ice is melting at a rapid rate, leaving the polar bear with limited areas to breed and hunt. The situation has already become so grave in a short period of time that the polar bear's survival is now in jeopardy. The polar bear is now under consideration as a threatened and endangered species.

Voluntary adaptation now will help necessary adaptation in the future. *(Nature's Images)*

Like the polar bear, many of the world's ecosystems are stressed by several types of disturbances such as pollution, fragmentation (isolation of habitat), and invasion of exotic species. These factors coupled with global warming are likely to affect ecosystems' natural resiliency and prevent them from being able to adapt long term.

As far as human adaptability, the Pew Center advises that some adaptation will involve the gradual evolution of present trends; other adaptations may come as unexpected surprises. Changes will involve sociopolitical, technological, economic, and cultural aspects.

Because of the reality that populations are increasing, more people live in coastal areas, and more people live in floodplains and in drought-prone areas, adaptation measures will be required as climate changes. Fortunately, however, technology has developed to a point that there are better means today to successfully respond to climate change than there were in the past. For example, agricultural practices have evolved to the point that most crop species have been able to be translocated thousands of miles from their regions of origin by resourceful farmers.

The polar bear faces extinction as global warming steadily destroys its habitat. As the Arctic ice retreats, polar bears are negatively affected. Lack of ice takes away valuable hunting ground and migration corridors. *(Publitek, Inc.)*

A critical key to success is reactive adaptation; how willing will populations be to permanently change behaviors in order to adapt to changing climates and environmental conditions? Topics that populations will have to address encompass issues such as:

- rationed water,
- changes in water use habits,
- changes in crop type,
- resource conservation plans,
- mandatory use of renewable energy, and
- restricted transportation types.

Based on a study conducted by the Pew Center, waiting to act until change has occurred can be more costly than making forward-looking responses that anticipate climate change, especially with coastal and floodplain development. A "wait-and-see" approach would be unwise with regard to ecosystem impacts. According to the Pew Center, "Proactive adaptation, unlike reactive adaptation, is forward-looking and takes into account the inherent uncertainties associated with anticipating change. Successful proactive adaptation strategies are flexile; they are designed to be flexible under a wide variety of climate conditions."

An extremely important note in adaptation that cannot be overlooked is government influence and public policy. Governments have a strong influence over the magnitude and distribution of climate change impacts and public preparedness. When climate and environmental disasters occur, it is usually government institutions that provide the necessary funding, develop the technologies, management systems, and support programs to minimize the occurrence of a repeat situation. A well-known example of this is the dust bowl that occurred in the midwestern United States in the 1930s. It was through the efforts of the U.S. government that conservation efforts were started in order to properly manage the nation's soil and agriculture in order to prevent a repeat disaster of that nature.

In view of global warming today and the already unstoppable effects in the future, adaptation and mitigation are necessary (and complementary concepts). Adaptation is a key requirement in order to

Public education is important in order to stop global warming—the better educated the public is, the sooner efficient global solutions will be reached. *(Nature's Images)*

lessen future damage. It is important to understand, however, that even though society will have to adapt, losses suffered will be inevitable and certain geographical areas will experience more extreme losses than others, particularly the developing countries.

According to a report that appeared in *National Public Radio News* (*NPR*) in December 2007, Australia is already taking proactive adaptation measures with its drinking water. Over the past 10 years, Perth, Australia, has been faced with drought conditions. Malcolm Turnbull, Australia's minister for the environment and water resources, says, "Over the past 10 years or so, the city has seen a 21 percent decline in rainfall, but the stream flow into dams—the actual amount running into storage—has dropped about 65 percent. We've seen similar declines in stream flow, though not quite so dramatic across southern Australia."

He further refers to Perth as Australia's "canary in the climate change coal mine"—a city scrambling to find other sources of water for a growing population. Due to the population growth in Perth and resource demands, the city's water resources are currently under great demand and pressure. Because of it, Western Australia Water Corp. has turned to the Indian Ocean to help solve the problem.

In order to adapt to the accelerating water demands, the Kwinana Desalination Plant was constructed—the first desalination plant in Australia. The plant is currently producing nearly 40 million gallons (151 million liters) of drinking water a day, equal to about 20 percent of Perth's daily consumption rate. Simon McKay, the project manager, says the desalination process from seawater to tap water only takes about 30 minutes. He also points out that in the past, desalination plants have been extremely expensive (the Middle East has used them for decades), but new technology has made them cheaper and more efficient.

Unfortunately, they still consume a large amount of energy. For this reason, the operation in Australia runs on wind energy, generated from 48 wind turbines, each as tall as a 15-story building at the Emu Downs Wind Farm.

Kerry Roberts, Emu Down's general manager, says, "If you look at the combined output of the wind farm at maximum wind speeds—24–28 miles per hour (39–45 km/hr)—you're looking at an output of close to 80 megawatts. That's enough power to run Perth's desalination plant 160 miles (257 km) to the south.

Gary Crisp of Western Australia Water Corp. says, "I predict that desalination will account for at least half of Perth's water in the next 30 years."

While it is understood today that both mitigation and adaptation must occur simultaneously, each country will be faced with different issues to resolve and overcome. This issue has been a major reason for the lack of universal cooperation among nations, specifically, a major stumbling block in getting the United States to ratify the Kyoto

Protocol, committing the country to put a mandatory curb on greenhouse gas emissions.

President George W. Bush consistently opposed any plan that included specific targets or binding commitments for reducing greenhouse gas emissions, which could have economic impacts, or which assign more responsibility for solving the problem of global warming to developed nations such as the United States than to developing economies such as China, India, and Brazil. Those were the two reasons President Bush gave for refusing to ratify the Kyoto Protocol along with the nations that did. As of February 2009, 181 nations have ratified the protocol.

On February 14, 2002, President Bush announced his proposal—an alternative plan aimed at reducing greenhouse gas emissions. Under the proposal, the United States would host and lead a series of talks that would include the nations that emit the most greenhouse gases—including large developing nations such as China and India. The aim of those negotiations would be to draft an agreement to replace the Kyoto Protocol when it expires in 2012 by setting "a long-term goal" to cut greenhouse gas emissions. Under the Bush plan, however, the goal would be "aspirational" and, therefore, voluntary. There would be no binding commitments, and each nation would be free to devise its own strategies for meeting the goal. Overall, the plan was met with criticism, with opponents inferring it did not have the "backbone" to make a real difference.

This topic is so controversial because it has such a moral dilemma attached to it—do wealthier nations help support developing nations by adopting an "open door" policy, or do they focus only on themselves? Which countries should take more responsibility and initiative—the wealthier ones or the biggest greenhouse gas contributors? These issues are very real and must be addressed in a practical, workable way before a solution can be obtained.

President Obama, in the current administration, has vowed to make global climate change a priority. Many environmental organizations, academic institutions, private and government researchers, the general American public, and foreign countries worldwide anxiously await to see what changes his administration will bring to the big picture of global warming and what solutions the United States is willing to commit to now. As temperatures continue to climb, time is of the essence.

Developing nations may face different issues of climate change than do developed nations. Varying geographic locations will also experience differing ranges of climate change. The only way to manage this overwhelming issue is for nations to work together in a global effort to control greenhouse gas emissions, understanding universal cause-and-effect relationships. Without international cooperation, there is little hope of stopping the problem before it is too late.

CHRONOLOGY

ca. 1400–1850 Little Ice Age covers the Earth with record cold, large glaciers, and snow. There is widespread disease, starvation, and death.

1800–70 The levels of CO_2 in the atmosphere are 290 ppm.

1824 Jean-Baptiste Joseph Fourier, a French mathematician and physicist, calculates that the Earth would be much colder without its protective atmosphere.

1827 Jean-Baptiste Joseph Fourier presents his theory about the Earth's warming. At this time many believe warming is a positive thing.

1859 John Tyndall, an Irish physicist, discovers that some gases exist in the atmosphere that block infrared radiation. He presents the concept that changes in the concentration of atmospheric gases could cause the climate to change.

1894 Beginning of the industrial pollution of the environment.

1913–14 Svante Arrhenius discovers the greenhouse effect and predicts that the Earth's atmosphere will continue to warm. He predicts that the atmosphere will not reach dangerous levels for thousands of years, so his theory is not received with any urgency.

1920–25 Texas and the Persian Gulf bring productive oil wells into operation, which begins the world's dependency on a relatively inexpensive form of energy.

1934 The worst dust storm of the dust bowl occurs in the United States on what historians would later call Black Sunday. Dust storms are a product of drought and soil erosion.

1945 The U.S. Office of Naval Research begins supporting many fields of science, including those that deal with climate change issues.

1949–50 Guy S. Callendar, a British steam engineer and inventor, propounds the theory that the greenhouse effect is linked to human actions and will cause problems. No one takes him too seriously, but scientists do begin to develop new ways to measure climate.

1950–70 Technological developments enable increased awareness about global warming and the enhanced greenhouse effect. Studies confirm a steadily rising CO_2 level. The public begins to notice and becomes concerned with air pollution issues.

1958 U.S. scientist Charles David Keeling of the Scripps Institution of Oceanography detects a yearly rise in atmospheric CO_2. He begins collecting continuous CO_2 readings at an observatory on Mauna Loa, Hawaii. The results became known as the famous Keeling Curve.

1963 Studies show that water vapor plays a significant part in making the climate sensitive to changes in CO_2 levels.

1968 Studies reveal the potential collapse of the Antarctic ice sheet, which would raise sea levels to dangerous heights, causing damage to places worldwide.

1972 Studies with ice cores reveal large climate shifts in the past.

1974 Significant drought and other unusual weather phenomenon over the past two years cause increased concern about climate change not only among scientists but with the public as a whole.

1976 Deforestation and other impacts on the ecosystem start to receive attention as major issues in the future of the world's climate.

1977 The scientific community begins focusing on global warming as a serious threat needing to be addressed within the next century.

1979 The World Climate Research Programme is launched to coordinate international research on global warming and climate change.

1982 Greenland ice cores show significant temperature oscillations over the past century.

1983 The greenhouse effect and related issues get pushed into the political arena through reports from the U.S. National Academy of Sciences and the Environmental Protection Agency.

1984–90 The media begins to make global warming and its enhanced greenhouse effect a common topic among Americans. Many critics emerge.

1987 An ice core from Antarctica analyzed by French and Russian scientists reveals an extremely close correlation between CO_2 and temperature going back more than 100,000 years.

1988 The United Nations set up a scientific authority to review the evidence on global warming. It is called the Intergovernmental Panel on Climate Change (IPCC) and consists of 2,500 scientists from countries around the world.

1989 The first IPCC report says that levels of human-made greenhouse gases are steadily increasing in the atmosphere and predicts that they will cause global warming.

1990 An appeal signed by 49 Nobel prizewinners and 700 members of the National Academy of Sciences states, "There is broad agreement within the scientific community that amplification of the Earth's natural greenhouse effect by the buildup of various gases introduced by human activity has the potential to produce dramatic changes in climate . . . Only by taking action now can we insure that future generations will not be put at risk."

1992 The United Nations Conference on Environment and Development (UNCED), known informally as the Earth Summit, begins on June 3 in Rio de Janeiro, Brazil. It results in the United Nations Framework Convention on Climate Change, Agenda 21, the Rio Declaration on Environment and Development Statement of Forest Principles, and the United Nations Convention on Biological Diversity.

1993 Greenland ice cores suggest that significant climate change can occur within one decade.

1995 The second IPCC report is issued and concludes there is a human-caused component to the greenhouse effect warming. The consensus is that serious warming is likely in the coming century. Reports on the breaking up of Antarctic ice sheets and other signs of warming in the polar regions are now beginning to catch the public's attention.

1997 The third conference of the parties to the Framework Convention on Climate Change is held in Kyoto, Japan. Adopted on December 11, a document called the Kyoto Protocol commits its signatories to reduce emissions of greenhouse gases.

2000 Climatologists label the 1990s the hottest decade on record.

2001 The IPPC's third report states that the evidence for anthropogenic global warming is incontrovertible, but that its effects on climate are still difficult to pin down. President Bush declares scientific uncertainty too great to justify Kyoto Protocol's targets.

The United States Global Change Research Program releases the findings of the National Assessment of the Potential Consequences of Climate Variability and Change. The assessment finds that temperatures in the United States will rise by 5 to 9°F (3–5°C) over the next century and predicts increases in both very wet (flooding) and very dry (drought) conditions. Many ecosystems are vulnerable to climate change. Water supply for human consumption and irrigation is at risk due to increased probability of drought, reduced snow pack, and increased risk of flooding. Sea-level rise and storm surges will most likely damage coastal infrastructure.

2002 Second hottest year on record.

Heavy rains cause disastrous flooding in Central Europe leading to more than 100 deaths and more than $30 billion in damage. Extreme drought in many parts of the world (Africa, India,

Australia, and the United States) results in thousands of deaths and significant crop damage. President Bush calls for 10 more years of research on climate change to clear up remaining uncertainties and proposes only voluntary measures to mitigate climate change until 2012.

2003 U.S. senators John McCain and Joseph Lieberman introduce a bipartisan bill to reduce emissions of greenhouse gases nationwide via a greenhouse gas emission cap and trade program.

Scientific observations raise concern that the collapse of ice sheets in Antarctica and Greenland can raise sea levels faster than previously thought.

A deadly summer heat wave in Europe convinces many in Europe of the urgency of controlling global warming but does not equally capture the attention of those living in the United States.

International Energy Agency (IEA) identifies China as the world's second largest carbon emitter because of their increased use of fossil fuels.

The level of CO_2 in the atmosphere reaches 382 ppm.

2004 Books and movies feature global warming.

2005 Kyoto Protocol takes effect on February 16. In addition, global warming is a topic at the G8 summit in Gleneagles, Scotland, where country leaders in attendance recognize climate change as a serious, long-term challenge.

Hurricane Katrina forces the U.S. public to face the issue of global warming.

2006 Former U.S. vice president Al Gore's *An Inconvenient Truth* draws attention to global warming in the United States.

Sir Nicholas Stern, former World Bank economist, reports that global warming will cost up to 20 percent of worldwide gross domestic product if nothing is done about it now.

2007 IPCC's fourth assessment report says glacial shrinkage, ice loss, and permafrost retreat are all signs that climate change is underway now. They predict a higher risk of drought, floods, and more powerful storms during the next 100 years. As a result, hunger, homelessness, and disease will increase. The atmosphere may warm 1.8 to 4.0°C and sea levels may rise 7 to 23 inches (18 to 59 cm) by the year 2100.

Al Gore and the IPCC share the Nobel Peace Prize for their efforts to bring the critical issues of global warming to the world's attention.

2008 The price of oil reached and surpassed $100 per barrel, leaving some countries paying more than $10 per gallon.

Energy Star appliance sales have nearly doubled. Energy Star is a U.S. government-backed program helping businesses and individuals protect the environment through superior energy efficiency.

U.S. wind energy capacity reaches 10,000 megawatts, which is enough to power 2.5 million homes.

2009 President Obama takes office and vows to address the issue of global warming and climate change by allowing individual states to move forward in controlling greenhouse gas emissions. As a result, American automakers can prepare for the future and build cars of tomorrow and reduce the country's dependence on foreign oil. Perhaps these measures will help restore national security and the health of the planet, and the U.S. government will no longer ignore the scientific facts.

The year 2009 will be a crucial year in the effort to address climate change. The meeting on December 7–18 in Copenhagen, Denmark, of the UN Climate Change Conference promises to shape an effective response to climate change. The snapping of an ice bridge in April 2009 linking the Wilkins Ice Shelf (the size of Jamaica) to Antarctic islands could cause the ice shelf to break away, the latest indication that there is no time to lose in addressing global warming.

APPENDIX

THE TOP 20 CARBON DIOXIDE EMITTERS

The world's countries contribute different amounts of heat-trapping gases to the atmosphere. Carbon emissions from all sources of fossil fuel burning for a maximum period from 1751 to 2004 (or as long as a record is available) are shown. The following table lists the 20 countries with the highest carbon emissions.

The Top 20 Carbon Dioxide Emitters

COUNTRY	TOTAL EMISSIONS (1,000 TONS TO C)	PER CAPITAL EMISSIONS (TONS/CAPITAL)	PER CAPITA EMISSIONS (RANK)
1. United States	1,650,020	5.61	(9)
2. China (mainland)	1,366,554	1.05	(92)
3. Russian Federation	415,951	2.89	(28)
4. India	366,301	0.34	(129)
5. Japan	343,117	2.69	(33)
6. Germany	220,596	2.67	(36)
7. Canada	174,401	5.46	(10)
8. United Kingdom	160,179	2.67	(37)
9. Republic of Korea	127,007	2.64	(39)
10. Italy	122,726	2.12	(50)
11. Mexico	119,473	1.14	(84)

COUNTRY	TOTAL EMISSIONS (1,000 TONS TO C)	PER CAPITAL EMISSIONS (TONS/CAPITAL)	PER CAPITA EMISSIONS (RANK)
12. South Africa	119,203	2.68	(34)
13. Iran	118,259	1.76	(63)
14. Indonesia	103,170	0.47	(121)
15. France/Monaco	101,927	1.64	(66)
16. Brazil	90,499	0.50	(118)
17. Spain	90,145	2.08	(52)
18. Ukraine	90,020	1.90	(56)
19. Australia	89,125	4.41	(13)
20. Saudi Arabia	84,116	3.71	(18)

Source: Marland, G., T. A. Boden, R. J. Andres 2004. Global, Regional, and National CO_2 Emissions. In Trends: A Compendium of Data on Global Change. Carbon Dioxide Information Analysis Center, Oak Ridge National Laboratory, U.S. Department of Energy, Oak Ridge, Tennessee.

Note: The above data was collected in 2004. In a report that appeared in the *New York Times* on June 14, 2008, China was identified as the largest emitter of CO_2 in the world. The report was released by the Netherlands Environmental Assessment Agency and found that in 2007 China's emissions were 14 percent higher than those of the United States. In the previous year's annual study, the researchers found for the first time that China had become the world's leading emitter, with carbon emissions 7 percent higher by volume than the United States in 2006.

GLOSSARY

adaptation an adjustment in natural or human systems to a new or changing environment. Adaptation to climate change refers to adjustment in natural or human systems in response to actual or expected climatic changes.

aerosols tiny bits of liquid or solid matter suspended in air. They come from natural sources such as erupting volcanoes and from waste gases emitted from automobiles, factories, and power plants. By reflecting sunlight, aerosols cool the climate and offset some of the warming caused by greenhouse gases.

albedo the relative reflectivity of a surface. A surface with high albedo reflects most of the light that shines on it and absorbs very little energy; a surface with a low albedo absorbs most of the light energy that shines on it and reflects very little.

anthropogenic made by people or resulting from human activities. This term is usually used in the context of emissions that are produced as a result of human activities.

atmosphere the thin layer of gases that surround the Earth and allow living organisms to breathe. It reaches 400 miles (644 km) above the surface, but 80 percent is concentrated in the **troposphere**—the lower seven miles (11 km) above the Earth's surface.

biodiversity different plant and animal species.

biomass plant material that can be used for fuel.

bleaching (coral) the loss of algae from corals that causes the corals to turn white. This is one of the results of **global warming** and signifies a die-off of unhealthy coral.

carbon a naturally abundant nonmetallic element that occurs in many inorganic and in all organic compounds.

carbon cycle the biogeochemical cycle by which carbon is exchanged among the biosphere, pedosphere, geosphere, hydrosphere, and atmosphere of the Earth.

carbon dioxide a colorless, odorless gas that passes out of the lungs during respiration. It is the primary **greenhouse gas** and causes the greatest amount of **global warming.**

carbon sink an area where large quantities of **carbon** are built up in the wood of trees, in calcium carbonate rocks, in animal species, in the ocean, or any other place where carbon is stored. These places act as a reservoir, keeping carbon out of the atmosphere.

climate the usual pattern of weather that is averaged over a long period of time.

climate model a quantitative way of representing the interactions of the atmosphere, oceans, land surface, and ice. Models can range from relatively simple to extremely complicated.

climatologist a scientist who studies the climate.

concentration the amount of a component in a given area or volume. In **global warming,** it is a measurement of how much of a particular gas is in the atmosphere compared to all of the gases in the atmosphere.

condense the process that changes a gas into a liquid.

deforestation the large-scale cutting of trees from a forested area, often leaving large areas bare and susceptible to erosion.

desertification the process that turns an area into a desert.

ecological the protection of the air, water, and other natural resources from pollution or its effects. It is the practice of good environmentalism.

ecosystem a community of interacting organisms and their physical environment.

electromagnetic spectrum the entire continuous spectrum of all forms of electromagnetic radiation, from gamma rays to long radio waves.

El Niño a cyclic weather event in which the waters of the eastern Pacific Ocean off the coast of South America become much warmer than normal and disturb weather patterns across the region. Its full name is El Niño Southern Oscillation (ENSO). Every few years, the temperature of the western Pacific rises several degrees above that of waters to the east. The warmer water moves eastward, causing shifts in ocean currents, jet-stream winds, and weather in both the Northern and Southern Hemispheres.

emissions the release of a substance (usually a gas when referring to the subject of climate change) into the atmosphere.

estuary the widened tidal mouth of a river valley where freshwater comes into contact with seawater and where tidal effects are evident.

evaporation the process by which a liquid, such as water, is changed to a gas.

evapotranspiration the transfer of moisture from the Earth to the atmosphere by evaporation of water and transpiration from plants.

feedback a change caused by a process that, in turn, may influence that process. Some changes caused by **global warming** may hasten the process of warming (positive feedback); some may slow warming (negative feedback).

food chain the transfer of food energy from producer (plant) to consumer (animal) to decomposer (insect, fungus, etc.).

forb an herbaceous plant that is not a grass.

forcings mechanisms that disrupt the global energy balance between incoming energy from the Sun and outgoing heat from the Earth. By altering the global energy balance, such mechanisms "force" the climate to change. Today, **anthropogenic greenhouse gases** added to the atmosphere are forcing climate to behave as it is.

fossil fuel an energy source made from coal, oil, or natural gas. The burning of fossil fuels are one of the chief causes of **global warming.**

glacier a mass of ice formed by the buildup of snow over hundreds and thousands of years.

global warming an increase in the temperature of the Earth's atmosphere, caused by the buildup of greenhouse gases. This is also referred to as the "enhanced greenhouse effect" caused by humans.

global warming potential (GWP) a measure of how much a given mass of **greenhouse gas** is estimated to contribute to **global warming.** It is a relative scale, which compares the gas in question to that of the same mass of carbon dioxide, whose GWP is equal to 1.

Great Ocean Conveyor Belt a global current system in the ocean that transports heat from one area to another.

greenhouse effect the natural trapping of heat energy by gases present in the atmosphere, such as **carbon dioxide, methane,** and water vapor. The trapped heat is then emitted as heat back to the Earth.

greenhouse gas a gas that traps heat in the atmosphere and keeps the Earth warm enough to allow life to exist.

hydrologic cycle the natural sequence through which water passes into the atmosphere as water vapor, precipitates to the Earth in liquid or solid form, and ultimately returns to the atmosphere through evaporation.

Industrial Revolution the period during which industry developed rapidly as a result of advances in technology. This took place in Britain during the late 18th and early 19th centuries.

IPCC Intergovernmental Panel on Climate Change. This is an organization consisting of 2,500 scientists that assesses information in the scientific and technical literature related to the issue of climate change. The United Nations Environment Programme and the World Meteorological Organization established the IPCC jointly in 1988.

land use the management practice of a certain land cover type. Land use may be such things as forest, arable land, grassland, urban land, and wilderness.

land-use change an alteration of the management practice on a certain land-cover type. Land-use changes may influence climate systems because they affect evapotranspiration and sources and sinks of greenhouse gases. An example of land-use change is removing a forest to build a city.

methane a colorless, odorless, flammable gas that is the major ingredient of natural gas. Methane is produced wherever decay occurs and little or no oxygen is present.

moldboard plowing plowing using a curved metal plate in a plow that turns over the earth from the furrow.

niche an environment in which an organism could successfully survive and reproduce in the absence of competition.

nitrogen as a gas, nitrogen takes up 80 percent of the volume of the Earth's atmosphere. It is also an element in substances such as fertilizer.

nitrous oxide a heat-absorbing gas in the Earth's atmosphere. Nitrous oxide is emitted from nitrogen-based fertilizers.

ozone a molecule that consists of three oxygen atoms. Ozone is present in small amounts in the Earth's atmosphere at 14 to 19 miles (23–31km) above the Earth's surface. A layer of ozone makes life possible by shielding the Earth's surface from most harmful ultraviolet rays. In the lower atmosphere, ozone emitted from auto exhausts and factories is an air pollutant.

parts per million (ppm) the number of parts of a chemical found in 1 million parts of a particular gas, liquid, or solid.

permafrost permanently frozen ground in the Arctic. As global warming increases, this ground is melting.

photosynthesis the process by which plants make food using light energy, **carbon dioxide,** and water.

protocol the terms of a treaty that have been agreed to and signed by all parties.

radiation the particles or waves of energy.

renewable something that can be replaced or regrown, such as trees, or a source of energy that never runs out, such as solar energy, wind energy, or geothermal energy.

resources the raw materials from the Earth that are used by humans to make useful things.

satellite any small object that orbits a larger one. Artificial satellites carry instruments for scientific study and communication. Imagery taken from satellites is used to monitor aspects of **global warming** such as glacier retreat, ice cap melting, **desertification,** erosion, hurricane damage, and flooding. Sea surface temperatures and measurements are also obtained from man-made satellites in orbit around the Earth.

simulation a computer model of a process that is based on actual facts. The model attempts to mimic, or replicate, actual physical processes.

symbiotic a relationship where two dissimilar organisms share a mutually beneficial relationship.

temperate an area that has a mild climate and different seasons.

thermal something that relates to heat.

trade winds winds that blow steadily from east to west and toward the equator. The trade winds are caused by hot air rising at the equator, with cool air moving in to take its place from the north and from the south. The winds are deflected westward because of the Earth's west-to-east rotation.

tropical a region that is hot and often wet (humid). These areas are located around the Earth's equator.

troposphere the bottom layer of the atmosphere, rising from sea level up to an average of about 7.5 miles (12 km).

tundra a vast treeless plain in the Arctic with a marshy surface covering a **permafrost** layer.

weather the conditions of the atmosphere at a particular time and place. Weather includes such measurements as temperature, precipitation, air pressure, and wind speed and direction.

FURTHER RESOURCES

JOURNAL ARTICLES

Alley, Richard B. "Abrupt Climate Change." *Scientific American* 291, no. 5 (November 2004): 62–69. This article discusses the Great Ocean Conveyor Belt current and how global warming could shut it down and trigger another ice age.

Bagla, Pallava. "Melting Himalayan Glaciers May Doom Towns." *National Geographic News* (May 7, 2002): Available online. URL: http://news.nationalgeographic.com/news/2002/05/0501_020502_himalaya.html. Accessed August 12, 2009. This article discusses the overflow of lakes from melting glaciers and the hazards they pose to nearby villages.

Bartleman, Bill. "Partnership to Test Underground Storage of Carbon Dioxide." *Paducah Sun,* 15 July 2008. Available online. URL: http://www.istockanalyst.com/article/viewiStockNews/articleid/2393941. Accessed April 18, 2008. This article discusses the feasibility of storing carbon dioxide underground at a site in Kentucky.

Beever, Erik A., David A. Pyke, Jeanne C. Chambers, Fred Landau, and Stanley D. Smith. "Monitoring Temporal Change in Riparian Vegetation of Great Basin National Park." *Western North American Naturalist* 65, no. 3 (July 2005): 382–402. This article discusses climate change and the effects on the migration of vegetation.

Biello, David. "Out of Sight: Out of Clime: Burying Carbon in a Vault of Sea and Rock." *Scientific American News* (July 14, 2008): Available online. URL: http://www.sciam.com/article.cfm?id=undersea-carbon-capture-and-storage. Accessed August 12, 2009. This article discusses the process of carbon sequestration into the Pacific Northwest seafloor.

———. "Climate Change Science Moves from Proof to Prevention." *Scientific American News* (February 1, 2007): Available online.

URL: http://www.sciam.com/article.cfm?id=climate-change-science-mo. Accessed August 12, 2009. This article focuses on the status of global warming and actions that need to be taken now in response.

Britt, Robert Roy. "Surprising Side Effects of Global Warming." *LiveScience.* Posted December 22, 2004. Available online. URL: http://www.livescience.com/environment/041222_permafrost.html. Accessed April 3, 2008. This article discusses the effects of global warming on the world's zones of permafrost.

———. "Insurance Company Warns of Global Warming's Costs." *LiveScience.* Posted November 1, 2005. Available online. URL: http://www.livescience.com/environment/051101_insurance_warming.html. Accessed October 3, 2008. This discusses the economic liabilities of costly damage as a result of global warming.

———. "Caution: Global Warming May Be Hazardous to Your Health." *LiveScience.* Posted February, 21, 2005. Available online. URL: http://www.livescience.com/environment/050221_warming_health.html. Accessed August 2, 2008. This article discusses health effects from contaminated water supplies and other health effects from global warming.

———. "The Irony of Global Warming: More Rain Less Water." *LiveScience.* Posted November 16, 2005. Available online. URL: http://www.livescience.com/environment/051116_water_shortage.html. Accessed December 10, 2008. This article discusses the plight of freshwater reserves and future available drinking water.

Charles, Dan. "Will a Warmer World Have Enough Food?" *NPR* (December 14, 2007): Available online. URL: http://www.npr.org/template/story/story.php?storyId=15737145. Accessed December 10, 2008. This article discusses changing rainfall patterns and its effects on agriculture.

Danielson, Stentor. "Everest Melting? High Signs of Climate Change." *National Geographic News* (June 5, 2002): Available online. URL: news.nationalgeographic.com/news/2002/06/0605_

020604_everestclimate.html. Accessed March 3, 2008. This article discusses the effects of global warming on glacial retreat.

Doughton, Sandi. "Ocean Could Be the Place to Store Fossil Fuel Mess." *Seattle Times*, 15 July 2008. Available online. URL: http://seattletimes.nwsource.com/html/nationworld/2008051806_carbonstorage15m.html. Accessed May 5, 2008. This article discusses mitigation options for long-term carbon storage in the world's oceans.

Eilperin, Juliet. "Antarctic Ice Sheet Is Melting Rapidly." *Washington Post*, 3 March 2006. Available online. URL:. http://www.washingtonpost.com/wp-dyn/content/article/2006/03/02/AR2006030201712_pf.html. Accessed August 12, 2009. This article discusses the ice mass melting in Antarctica.

Gurney, Kevin. "Revolutionary CO_2 Maps Zoom in on Greenhouse Gas Sources." *Purdue University News*, 7 April 2008. Available online. URL: http://news.uns.purdue.edu/x/2008a/080407GurneyVulcan.html. Accessed August 12, 2009. This article presents a newly developed modeling technique to illustrate principal areas of carbon dioxide emissions in the United States.

Goldberg, David. "A Deep-Sea Storage Locker for Greenhouse Gas." *Discover* (15 July 2008): Available online. URL: http://blogs.discovermagazine.com/80beats/2008/07/15/a-deep-sea-storage-locker-for-greenhouse-gas/. Accessed September 15, 2008. This article discusses sea floor carbon sequestration options off the coast of the Pacific Northwest.

Goudarzi, Sara. "Allergies Getting Worse Due to Global Warming." *LiveScience.* Posted November 22, 2005. Available online. URL: http://www.livescience.com/health/051122_allergy_rise.html. Accessed August 12, 2009. This article discusses the effects higher temperatures are having on chronic allergies in the United States.

Handwerk, Brian. "Global Warming Could Cause Mass Extinctions By 2050, Study Says." *National Geographic News* (April 12, 2006): Available online. URL: http://news.nationalgeographic.com/news/pf/1930686.html. Accessed November 23, 2008. This article presents evidence for extinctions worldwide in all habitats due to a continuing rise in CO_2.

Harmon, M. E., W. K. Ferrell, and J. K. Franklin. "Effects on Carbon Storage of Conversion of Old-Growth Forests to Young Forests." *Science* 247, no. 4,943 (February 1990): 699–702. This article discusses how much carbon newer, faster-growing trees can hold in relation to older slower-growing trees.

Hennessey, Kathleen. "Dairy Air: Scientists Measure Cow Gas." *LiveScience.* Posted July 27, 2005. Available online. URL: http://www.livescience.com/animals/ap_050727_cow_gas.html. Accessed August 12, 2009. This article discusses the contribution of greenhouse gases from the grazing industry.

Hopkins, Philip. "Global Warming Cyclical, Says Climate Expert." *The Age Company* (June 13, 2005): Available online. URL: http://www.theage.com.au/news/Science/Global-warming-cyclical-says-climate-expert/2005/06/ 12/1118514924793.html. Accessed April 6, 2008. This review examines global warming from the aspect that it is just part of the general cyclic climatic cycles the Earth naturally goes through.

Houghton, R. A., J. L. Hackler, and K. T. Lawrence. "The U.S. Carbon Budget: Contributions from Land-Use Change." *Science* 285, no. 5,427 (July 1999): 574–578. This review takes a close look at how the qualify of the landscape is affected by changes in land-use practices.

Kloor, K. "Restoration Ecology: Returning America's Forests to Their 'Natural' Roots." *Science* 287, no. 5,453 (January 2000): 573–575.

Maffly, Brian, and Judy Fahys. "Setting Trap for Global Warming." *Salt Lake Tribune,* 16 November 2007. Available online. URL: http://www.utahmining.org/07nov.htm. Accessed August 12, 2009. This article discusses a new technology to enable carbon sequestration to offset fossil fuel pollution produced by power plants in Utah.

Marchetti, C. "On Geoengineering and the CO_2 Problem." *Climate Change,* no. 1 (1977): 59–68. This article discusses carbon dioxide sequestration into the North Atlantic below the thermocline.

McCarl, Bruce A., and Uwe A. Schneider. "Greenhouse Gas Mitigation in U.S. Agriculture and Forestry." *Science* 294, no. 5,551 (December 21, 2001): 2,481–2,482. This article discusses carbon sequestration in forest and agricultural areas and the effects proper land management can have on the total amount of carbon stored in the vegetation and soils.

McClure, Robert. "Changing Wildlife Patterns Linked to Global Warming." *Seattle Post-Intelligencer Reporter,* 13 February 2002. Available online. URL: http://seattlepi.nwsource.com/national/58041_clim13.shtml. Accessed August 11, 2009. This article discusses shifts in habitats as a result of global warming.

Minarcek, Andrea. "Mount Kilimanjaro's Glacier Is Crumbling." *National Geographic News* (September 23, 2003): Available online. URL: news.nationalgeographic.com/news/2003/09/0923_030923_kilimanjaroglaciers.html. Accessed August 11, 2009. This article discusses the issue of glacial melt as a result of global warming.

MSNBC News Services. "Warming Expert: Only Decade Left to Act In." 14 September 2006. Available online. URL: http://www.msnbc.msn.com/id/14834318/. Accessed August 11, 2009. This report reviews a public presentation or James Hansen gave concerning the necessity to act soon in order to curtail the detrimental effects of global warming before it is too late.

New York Times. "The Scientists Speak." 20 November 2007. Available online. URL: http://www.nytimes.com/2007/11/20/opinion/20tue1.html. Accessed August 11, 2009. This article explores the global warming issue from the scientific viewpoint and provides recommendations on where solving the problem needs to go from here.

New York Times. "No Place to Hide." 20 January 2008. Available online. URL: http://www.nytimes.com/2008/01/20/opinion/20sun4.html. Accessed August 11, 2009. This article examines the worldwide effect of global warming and the importance of taking action now.

Norby, R. J., E. H. DeLucia, B. Gielen, C. Calfapietra, C. P. Glardina, J. S. King, J. Ledford, H. R. McCarthy, et al. "Forest Response to

Elevated CO_2 Is Conserved Across a Broad Range of Productivity." *Proceedings of the National Academy of Sciences* 102, no. 50 (2005): 18,052–18,056. This article examines forest habitats and their response to increased levels of CO_2 with global warming.

Noss, R. F. "Beyond Kyoto: Forest Management in a Time of Rapid Climate Change." *Conservation Biology* 15, no. 3 (2001): 578–590. This article examines issues concerning resources and quality of the environment.

O'Driscoll, Patrick. "Study Links Extended Wildfire Seasons to Global Warming." *USA Today,* 7 July 2006. Available online. URL: http://www.waterconserve.org/shared/reader/welcome.aspx?linkid=58066. Accessed December 30, 2008. This article correlates the number and size of large forest fires in the West with global warming.

Pickrell, John. "Global Warming May Boost Crop Yields, Study Says." *National Geographic News* (November 25, 2002): Available online. URL: http://news.nationalgeographic.com/news/2002/11/1122_021125_CropYields.html. Accessed August 11, 2009. This article explores the possibilities of increased yields in select areas due to increased levels of CO_2.

Revkin, Andrew C. "A New Middle Stance Emerges in Debate Over Climate." *New York Times,* 1 January 2007. Available online. URL: http://www.nytimes.com/2007/01/01/science/01climate.html. Accessed August 11, 2009. This article examines global warming from a new viewpoint, this one not of extremes.

———. "A Shift in the Debate Over Global Warming." *New York Times,* 6 April 2008. Available online. URL: http://www.pewclimate.org/node/5869. Accessed August 11, 2009. This article examines emission caps, energy conservation, and renewable resources to combat global warming.

———. "Strong Action Urged to Curb Warming." *New York Times,* 11 June 2008. Available online. URL: www.nytimes.com/2008/06/11/world/11climate.html?pagewanted=print. Accessed August 11, 2009. This article examines the activities of industrialized nations and their efforts to curb greenhouse gas emissions.

Roach, John. "Studies Measure Capacity of "Carbon Sinks." *National Geographic News* (June 21, 2001): Available online. URL: http://news.nationalgeographic.com/news/2001/06/0621_carbonsinks.html. Accessed August 11, 2009. This article presents a study to illustrate how much carbon is stored in sinks in the United States.

———. "Freshwater Runoff Into Arctic on the Rise, Scientists Say." *National Geographic News* (December 13, 2002): Available online. URL: news.nationalgeographic.com/news/2002/12/1212_021213_arcticrivers.html. Accessed August 11, 2009. This article examines the effects of freshwater on ocean currents.

———. "Antarctic Glaciers Surged after 1995 Ice-Shelf Collapse." *National Geographic News* (March 6, 2003): Available online. URL: news.nationalgeographic.com/news/2003/03/0306_030306_glaciersurge.html. Accessed August 11, 2009. This article discusses glacial surge episodes in connection with ice-shelf activity.

Rosenthal, Elisabeth, and Andrew C. Revkin. "Science Panel Calls Global Warming "Unequivocal." *New York Times,* 3 February 2007. Available online. URL: http://www.nytimes.com/2007/02/03/science/earth/03climate.html. Accessed August 11, 2009. This article provides a review of the IPCC's most current information on global warming.

Schimel, D. S., et al. "Recent Patterns and Mechanisms of Carbon Exchange by Terrestrial Ecosystems." *Nature* 414, (November 8, 2001): 169–172. This looks at how various members of an ecosystem store and sequests carbon.

Schirber, Michael. "Nature's Wrath: Global Death and Costs Swell." *LiveScience.* Posted November 1, 2004. Available online. URL: www.livescience.com. Accessed January 1, 2008. This article discusses the damage done to property and life by global warming.

Schlyer, Krista. "America's 10 Most Endangered National Wildlife Refuges." Defenders of Wildlife, 2007. Available online. URL: www.defenders.org. Accessed November 2, 2008. This article discusses the fragile ecosystems found within 10 U.S. wildlife refuges and what challenges they face for the future.

Schmid, Randolph E. "Global Warming Differences Resolved." *LiveScience*. Posted May 2, 2006. Available online. URL: http://www.livescience.com/environment/ap_060502_global_warming.html. Accessed January 1, 2009. This article presents the global warming debate as one that has a central focus because the key conflict of surface/atmosphere temperature discrepancies have been resolved.

Shah, Anup. "Climate Change and Global Warming." Horizon documentary on BBC, January 15, 2005. Available online. URL: http://www.globalissues.org/article/529/global-dimming. Accessed August 11, 2009. This program provides a good overview of the problem scientists and policy makers are up against.

Sullivan, Michael. "Australia Turns to Desalination Amid Water Shortage." *NPR* (December 14, 2007): Available online. URL: http://www.npr.org/templates/story/story.php?storyId=11134967. Accessed August 11, 2009. This article discusses action Australia is taking to adapt to water shortages by desalinating water supplies.

Sutherly, Ben. "Greenville Injection Project Could Have Global Implications." *Dayton Daily News,* 29 June 2008. Available online. URL: http://www.daytondailynews.com/n/content/oh/story/news/local/2008/06/28/ddn062908battellei nside.html. Accessed August 11, 2009. This article focuses on carbon sequestration issues and feasibility.

Than, Ker. "How Global Warming Is Changing the Wild Kingdom." *LiveScience*. Posted June 21, 2005. Available online. URL: www.livescience.com/environment/050621_warming_changes.html. Accessed August 11, 2009. This article discusses the current and future threats to wildlife habitats as a result of global warming.

Times of India. "Transforming Carbon Emissions." 14 July 2008. Available online. URL: http://en.newspeg.com/Transforming-carbon-emissions-5036947.html. Accessed January 1, 2009. This article discusses mineral sequestration for carbon.

Trivedi, Bijal P. "Antarctica Gives Mixed Signals on Warming." *National Geographic News* (January 25, 2002): Available online. URL: news.nationalgeographic.com/news/2002/01/0125_020125_ antarcticaclimate.html. Accessed August 11, 2009. This article discusses various reports of global warming research and their implications.

Wald, Matthew L. "Study Details How U.S. Could Cut 28% of Greenhouse Gases." *New York Times,* 30 November 2007. Available online. URL: http://royaldutchshellplc.com/2007/12/01/the-new-york-times-study-details-how-us-could-cut-28-of-greenhouse-gases/. Accessed August 11, 2009. This article offers several practical ways to be more energy efficient.

BOOKS

Martin, Parry L., Osvaldo F. Canziani, Jean P. Palutikof, Paul J. VanderLinden, and Clair E. Hanson. *IPCC Climate Change 2007: Impacts, Adaptation, and Vulnerability: Contribution of Working Group II to the Third Assessment Report of the Intergovernmental Panel on Climate Change.* Cambridge: Cambridge University Press, 2007.

Stork, N. E. "Measuring Global Biodiversity and Its Decline." In *Biodiversity: Understanding and Protecting our Natural Resources* by M. L. Reaka-Kudla, et al., 41–68. Washington D.C.: Joseph Henry Press, 1997. This book examines the changes in biodiversity and possibilities of extinction as a result of global warming.

WEB SITES

Global Warming

BBC: Climate Change. Sponsored by the BBC Weather Centre, UK. Available online. URL: www.bbc.co.uk/climate/. Accessed August 11, 2009. This Web site provides evidence, impacts, adaptation, policies and links about climate change.

Climate Ark home page. Sponsored by Ecological Internet. Available online. URL: www.climateark.org. Accessed August 11, 2009. This Web site promotes public policy that addresses global climate change through reduction in carbon and other emissions, energy conservation, alternative energy sources, and ending deforestation.

Climate Solutions home page. Sponsored by Atmosphere Alliance and Energy Outreach Center. Available online. URL: www.climatesolutions.org. Accessed August 11, 2009. This Web site offers practical solutions to global warming.

Environmental Defense Fund home page. Sponsored by Environmental Defense Fund. Available online. URL: www.environmentaldefense.org. Accessed August 11, 2009. A Web site from an organization started by a handful of environmental scientists in 1967 that provides quality information and helpful resources on understanding global warming and other crucial environmental issues.

Environmental Protection Agency home page. Sponsored by the U.S. Environmental Protection Agency. Available online. URL: www.epa.gov. Accessed August 11, 2009. This Web site provides information about EPA's efforts and programs to protect the environment. It offers a wide array of information on global warming.

European Environment Agency home page. Sponsored by the European Environment Agency in Copenhagen, Denmark. Available online. URL: www.eea.europa.eu/themes/climate. Accessed August 11, 2009. This Web site posts their reports on topics such as air quality, ozone depletion, and climate change.

Global Warming FAQs. Sponsored by the National Oceanic and Atmospheric Administration. Available online. URL: http://lwf.ncdc.noaa.gov/oa/climate/globalwarming.html. Accessed August 11, 2009. This Web site presents basic questions and answers about global warming.

Global Warming: Focus on the Future home page. Sponsored by EnviroLink. Available online. URL: www.enviroweb.org. Accessed August 11, 2009. This Web site offers statistics and photography of global warming topics.

HotEarth.Net home page. Sponsored by National Environmental Trust. Available online. URL: www.net.org/warming. Accessed August 11, 2009. This Web site features informational articles on the causes of global warming, its harmful effects, and solutions that could stop it.

Intergovernmental Panel on Climate Change (IPCC) home page. Sponsored by the World Meteorological Organization (WMO) and the United Nations Environment Programme (UNEP). Available online. URL: http://www.ipcc.ch/. Accessed August 11, 2009. This Web site offers current information on the science of global warming and recommendations on practical solutions and policy management.

NASA's Goddard Institute for Space Studies home page. Sponsored by the National Aeronautics and Space Administration. Available online. URL: www.giss.nasa.gov. Accessed August 11, 2009. This Web site provides a large database of information, research, and other resources.

NOAA's National Climatic Data Center home page. Sponsored by the National Oceanic and Atmospheric Administration. Available online. URL: www.ncdc.noaa.gov. Accessed October 26, 2007. This Web site offers a multitude of resources and information on climate, climate change, global warming.

Ozone Action home page. Sponsored by the Southeast Michigan Council of Governments. Available online. URL: www.semcog.org/OzoneAction_kids.aspx. Accessed August 11, 2009. This Web site provides information on air quality by focusing on ozone, the atmosphere, environmental issues, and related health issues.

Scientific American home page. Sponsored by Scientific American, Inc. Available online. URL: www.sciam.com. Accessed August 11, 2009. This organization offers an online magazine and often presents articles concerning climate change and global warming.

Tyndall Centre at University of East Anglia home page. Sponsored by the Tyndall Centre for Climate Change Research. Available online. URL: http://www.tyndall.ac.uk. Accessed August 11, 2009. This Web site offers information on climate change and is considered one of the leaders in United Kingdom research on global warming.

UK's Climatic Research Unit home page. Sponsored by the Climate Research Unit, School of Environmental Sciences, University of East Anglia, Norwich, United Kingdom. Available online. URL: www.

cru.uea.ac.uk/. Accessed August 11, 2009. This Web site aims to "improve scientific understanding of past climate history and its impact on humanity, the course and causes of climate change during the present century, and future prospects."

Union of Concerned Scientists home page. Sponsored by the Union of Concerned Scientists. Available online. URL: www.ucsusa.org. Accessed August 11, 2009. This Web site offers quality resource sections on global warming and ozone depletion.

United Nations Environment Programme-World Meteorological Organization home page. Sponsored by the United Nations Environment Programme. Available online. URL: www.gcrio.org/ipcc/qa/index.htm. Accessed August 11, 2009. This Web site contains an easy-to-follow question-and-answer presentation on global warming.

United Nations Framework Convention on Climate Change (UNFCCC) home page. Sponsored by the United Nations Framework Convention on Climate Change. Available online. URL: http://unfccc.int/2860.php. Accessed August 11, 2009. This Web site presents a spectrum on climate change information and policy.

United States Global Change Research Program home page. Sponsored by the U.S. Office of Science and Technology Policy, the Office of Management and Budget, and the Council on Environmental Quality. Available online. URL: www.globalchange.gov. Accessed August 11, 2009. This Web site provides information on the current research activities of national and international science programs that focus on global monitoring of climate and ecosystem issues.

World Wildlife Foundation Climate Change Campaign home page. Sponsored by the World Wildlife Fund. Available online. URL: www.worldwildlife.org/climate/. Accessed August 11, 2009. This Web site contains information on what various countries are doing, and not doing, to deal with global warming.

Greenhouse Gas Emissions

Energy Information Administration home page. Sponsored by the U.S. Department of Energy. Available online. URL: www.eia.doe.

gov/environment.html. Accessed August 11, 2009. This Web site lists official environmental energy-related emissions data and environmental analyses from the U.S. government. It contains U.S. carbon dioxide, methane, and nitrous oxide emissions data and other greenhouse reports.

World Resources Institute—Climate, Energy & Transport home page. Sponsored by the World Resources Institute. Available online. URL: www.wri.org/climate/publications.cfm. Accessed August 11, 2009. This Web site offers a collection of reports on global technology deployment to stabilize emissions, agriculture, and greenhouse gas mitigation, climate science discoveries, and renewable energy.

INDEX 

Italic page numbers indicate illustrations or maps. Page numbers followed by *c* denote entries in the chronology.

A

Academies of the Group of 8 Industrialized Countries 170
Adams, R. M. 66
adaptation 61, 68, 224–233, *227*
adaptive capacity 226
Adélie penguins 146
aerobic bacteria 55
afforestation 91
Africa 28, 86, 179–181
agricultural sequestration 43–45, *45*
agriculture
- air pollution and 82
- average temperature increase and 81
- biofuel production 79–80
- climate change response 61–86
- deforestation and 88, 90
- extreme events and 81
- and food supply 80–86
- global warming's effects on 60–61, 66–74
- Great Plains 195
- and greenhouse gases 60–86
- and greenhouse gas mitigation 74–79, *76, 77*
- midwestern U.S. 194
- Pacific Northwest 199
- precipitation changes 81–82
- and rising CO_2 concentrations 82–83, 85–86
- southeastern U.S. 190–191
- total greenhouse gas emissions 73–74

AgSTAR program 72–73
airlines 225
air pollution 58–59, 82
air quality regulations 69
Alaska 159–160, 165–166, 201–203, *206*
albedo 6, 15–16
alligators 157
alluvial fans 55
alpine ecosystems 198
alpine wildlife 150–151
Amazon region 68, 104, 107–108, 122
American alligators 157
American lobsters 157
American pikas 157
American robins 153, 156
ammonium 72
anaerobic decomposition 19
anaerobic digester systems 70–71
anaerobic lagoons 70
Andersons Marathon Ethanol LLC 53
Andres, Robert 59
Ångström, Knut 2
Antarctica 26, 146, 182–184, 204–205
Antarctic Ice Sheet 235*c*, 237*c*
Antarctic ice shelves 182–183
anthropogenic causes/effects of greenhouse gases 113–140
- Arrhenius's studies 3–4
- Callendar's studies 4
- carbon footprint 114–119, *119*
- cities at risk 128–138
- and climate change 18
- cultural losses 138–139
- early greenhouse effect studies 2
- economic impacts 139–140
- land-use change 119–122
- public lands 125–128

anthropogenic greenhouse warming. *See* enhanced greenhouse effect
Arctic 143–144, 146, 148, 181–182
Arctic sea ice melt 118, 202, 203
Arrhenius, Svante 2–4, 234*c*
Asia 131, 173–176
atmosphere 5, 9, 32
atmospheric lifetime of greenhouse gases 22
Australia 51, 180, 230–231
automobiles 38
Aziende Agricole Miranti 177

B

bacteria 20, 35, 69, 70
Baker, B. B. 66
bald eagles 164
Bangladesh 132
bark beetle 106, 178

barn swallow 153
basalt 49, 51, 56, 221
Bass, Brad 136
Bavaria, Germany 178
bees 177
Beever, Erik 151–152
Belize Barrier Reef 139
Beshear, Steve 50–51
Bhutan 174
bicarbonates 55
bighorn sheep 147
bio-bubble 69
biodiversity 89, 101–102, 108, 110–112
biofuels 79–80
biogas 71–72
biomass 35, 43, 55
biome 125
biosphere 31, 54
birds 153, 163, 164
black guillemots 156
Black Sunday 234*c*
bleaching (coral) 123, 157
Boreal Ecosystem-Atmosphere Study 96
boreal forest 36–37, 96–97, 109, 122
boron 86
Boston, Massachusetts 138
Bourdon, R. M. 66
Bowersox, Rick 50
Brazil 63
Brennan, Jean 167
Buddemieir, Robert 132
building heating/cooling 38
Burgelt, Jeff 167
Bush, George W. 49, 232, 238*c*
butterflies 163

C

Cai, Ming 121
calcium carbonate 38–39, 53
California 69, 105, 165
Callendar, Guy S. 4, 235*c*
Callendar Effect 4
cancer drugs 103
cap-and-trade programs 91–92, 168, 238*c*
Cape May National Wildlife Refuge 160–162
Cape May warbler 153
cap rock 50
carbonates 56
carbonate salts 53
carbon balance 17
carbon budget 94, 97
carbon capture and storage (CCS) 75, 77–79, 210–213
carbon credits 57
carbon cycle 30–32, *38*, 59, 120
carbon dioxide (CO_2)
  Arrhenius's studies 3–4
  Callendar's studies 4
  common sources 19
  and crop productivity 66–68, 82–86
  deforestation and 88
  and discovery of greenhouse effect 1, 2
  and enhanced greenhouse effect 10
  fertilizer manufacturing and 73
  and forest productivity 105
  global warming's effects on 54
  half-life 18
  ice core studies 236*c*
  increases (1950–1970) 235*c*
  and Keeling Curve 13–14
  levels, 1800–1870 234*c*
  levels, 2003 238*c*
  levels since 2000 117–118
  levels since Industrial Revolution 10, 13
  modeling of U.S. emissions 58–59
  and natural greenhouse effect 11
  and photosynthesis rates 62–63
  potency relative to other greenhouse gases 21–24
  sequestration. *See* carbon sequestration
carbon footprint 114–119, *119*
carbonic acid 53, 210
carbon labeling 116
carbon monoxide 35
carbon neutrality 116
carbon offsets 116
Carbon Sciences 56–57
carbon sequestration 3–59
  agricultural 43–45, *45*
  carbon cycle and 30–32, *38*
  carbon sinks and sources 32–39
  CCS 75, 77–79, 210–213
  forestry 45–49, *48*, 87–89
  geologic 49–53, 214–219, *216*
  mineral 53, 55
  and modeling of U.S. carbon emissions 58–59
  ocean 55–56, 219–223, *220, 222*
  potential drawbacks 57–58
  various methods 55–57
carbon sinks 31–39, 94–98, *95*, 209. *See also specific carbon sinks, e.g.:* forests

caribou 148, 203
catalytic converters 20
Cativela, J. P. 70
cattle 19, *65*. *See also* livestock
CCS. *See* carbon capture and storage
CDM (Clean Development Mechanism) 90–91
cement production 35, 38–39
Central America 184–186
Central Europe floods (2002) 237*c*
CFCs. *See* chlorofluorocarbons
CFP (Climate Friendly Parks) 125
chalk 51–52
Chamberlin, Thomas 2
Chang, Andrew 86
chemistry of greenhouse gases 18–21
Chicago heat wave (1995) 136, 194
China 49, 117, 129, 173, 238*c*
Chinook salmon 163
chlorofluorocarbons (CFCs) 20, 25–27, 32
CITES (Convention on International Trade in Endangered Species) 146
cities. *See* urban areas
citrus crops 62
clathrate hydrates 55
clay minerals 53
Clean Air Act (1990) 25
Clean Development Mechanism (CDM) 90–91
clear-cut logging 110
Climate Change Action Plan 73
Climate Friendly Parks (CFP) 125
climate models
  agricultural systems 61
  Alaska 201
  deforestation and 104
  forests and climate change 106–108
  global dimming and 29
  land use change 122
  midwestern U.S. 193
  radiative forcings 16
  western U.S. 197, 198
clouds 9, 68
coal 2
coal gasification 50
coal seams 50
Cohen, Jordan 19–20
combustion 35
commercial crops 63
Congo Basin Forest Partnership 120
Congress, U.S. 91, 145
coniferous forests 200
ConocoPhillips 50
conservation tillage 40, 43, 45, 75, *76*, 78
Convention on International Trade in Endangered Species (CITES) 146
coral reefs 149–150, 157, 180
corrosion 43
cows. *See* livestock
Craven Farms (Cloverdale, Oregon) 71
Crisp, Gary 231
crop productivity 63, 66–68, 82–86, 189–191, 194
Crow, Debra 53
cud 69
cultural losses, from anthropogenic climate change 138–139
Curtis, Peter S. 66–68
Czech Republic 139

D

Dairy CARES 70
decay, of biomass 11, 43
deforestation *47*, 87–112, *99*, *100*, 235*c*
  climate change impacts on forests 105–109, *107*
  and CO_2 emissions 19, 47
  forests as carbon sink 94–98, *95*
  forests' role in climate change 87–94
  impacts of 98–104, *99*, *100*
  land management after 109–112
  and removal of carbon sinks 33
  and surface temperatures 122
  and total greenhouse gas emissions 73
Delta Airlines 225
desalination 86, 231
desertification 81
developing countries
  agricultural greenhouse gas emissions 73–74
  crop productivity 83, *84*
  forest health 108
  heat waves in urban areas 132, 134
  tropic deforestation 99
DGVMs (Dynamic Global Vegetation Models) 106
Dicks, Norm 168
diffuse insolation 6
direct injection (ocean sequestration) 55, 56, 219
direct insolation 6
Doney, Scott 209
double cropping 194
drought
  1970s 235*c*
  2002 180, 237*c*–238*c*

and agriculture 62
American Southwest 65
Australia 180
effect on rangeland 65
global dimming and 28
and public lands 123
U.S. farming practices 74
and wildfires 35
drugs. *See* pharmaceuticals
"dry ice torpedoes" 219
Dubois, Jacques 140
ducks 146–147
dunite 55
dust bowl 234*c*
Dynamic Global Vegetation Models (DGVMs) 106

E

eagles 164
Earth Summit (Rio de Janeiro, 1992) 236*c*
economic impacts 101, 108, 126–127, 139–140
ecosystems 101, 198, 226–227
Edith's checkerspot butterflies 156
Eiffel Tower 139
electric power generation 37, 71, 79
electromagnetic spectrum 5–7
El Niño 123, 179, 184, 190, 200
Emu Downs Wind Farm 231
endangered species 57–58, 101, 110–112, 143–159
Endangered Species Act 144–146, 154–155
endemic species 101
Energy, U.S. Department of 53, 217
energy balance 3, 6–7, 16–18
energy crops 79
energy efficiency 225
energy production 37. *See also* fossil fuels
ENERGY STAR® program 37, 116, 239*c*
Englemann spruce 150
enhanced greenhouse effect 10–16, 235*c*
enhanced oil recovery 215
Environmental Quality Incentives Program (EQIP) 73
Epstein, Paul 140
erosion 66, 202
ethanol 53, 69, 79
Europe 172–178, 182, 238*c*
Everest, Mount 173
extinctions 110–112, 141, 146, 170
extreme weather events 41, 64–65, 81, 197

F

fallow periods 77–78
famine 28
Farnham Dome formation 218
Feddema, Johannes 122
Ferguson, Martin 51
fertilizers
from anaerobic digester systems 71
efficient use of 73, 78
and forest regrowth 109
manure 70
N_2O and 20, 41, 73
negative effects of 62
fish 149, 202
flatulence 19
flooding
and agriculture 62
Central Europe 237*c*
cultural losses from 138–139
Pacific Northwest 200
and sea-level rise 131
southeastern U.S. 190, 192
Florida panther *162*
fluorocarbons 20–21
food chain 109
food supply 80–86, *84*, 101–103, 108, 202–203. *See also* agriculture; crop productivity
forest ecosystems 57, 178
forest fires *36*
and CO_2 emissions 11, 19, 35
and ecosystem health 93–94
and public lands 127–128
forests
as carbon sink 32–33, 87–89, 94–98, *95*
climate change impacts on 105–109, *107*
midwestern U.S. 194
Pacific Northwest 200
role in climate change 87–94
southeastern U.S. 191–192
forest sequestration 45–49, *48*
fossil fuels
Arrhenius's studies 4
as carbon source 12, 19, 35, 37
in China 238*c*
Vulcan CO_2 modeling system and 58–59
Fourier, Jean-Baptiste-Joseph 2, 3, 234*c*
Framework Convention on Climate Change 236*c*, 237*c*
freshwater 181–182, 192

freshwater ecosystems 195
frosts 62
Frumhoff, Peter 101
fungi 35

G

G-8 countries 170, 238*c*
Gaffin, Stuart 132, 134–136
geologic sequestration 49–53, 212–219, *216*
Geoscience Australia 51
Germany 178
glacial flow 183
glacial melt/retreat 4, 123, *124,* 173, 174, 179
glacial surge 182–183
Glacier National Park 123
global dimming 27–29, *28*
global warming
  agriculture and 61–86, *84*
  around the world 171–203
  and boreal forest 97
  early focus on 235*c*
  effect on atmospheric CO_2 54
  effect on developing world 74
  effect on forests 105–109, *107*
  effect on oceanic CO_2 33–34
  effect on sequestration 41
  effect on wildlife 110–112, 141–144, *144,* 146–152, *147, 151*
  forests' role in 87–94
  Fourier's theories 234*c*
  general effects of 18, 204–208
  global dimming and 29
  land use change and 121–122
  media attention for 236*c*
  simple mitigation actions 148–149
  Vulcan system and 59
global warming potential (GWP) 21–25
Global Warming Wildlife Survival Act 168–170
Glyer, J. D. 66
Goldberg, David 49, 51, 52, 56, 221, 222
golden toads 157
Gore, Al 238*c,* 239*c*
Goulden, Mike 36, 37
Grand Teton National Park *125*
grazable land 63–64
Great Basin 151
Great Plains 82, 195–196
"green carbon" technology 56–57
greenhouse effect 1–29, *8,* 234*c*–236*c*
Greenland 236*c,* 237*c*
Greenland ice sheet 131
green-roof system 136
Group of 8 (G-8) 170, 238*c*
growing seasons, length of 62, 194
Gurney, Kevin 58, 59
GWP (global warming potential) 21–25

H

Hadley circulation 122
Hailstone National Wildlife Refuge 162–163
half-life of CO_2 18
Hall, Forrest G. 96
halocarbons 20
Hansen, James E. 10–11, 18, 59, 223
Hanson, J. D. 66
hardwoods 89
Harmon, M. E. 93
Harris, Charles 207
Hawaii 119
health insurance 139
heat islands 104, 122. *See also* urban heat island effect
heat waves 238*c*
Hellisheidi Geothermal Power Plant (Iceland) 52
HFCs (hydrofluorocarbons) 20–22
high-mountain species 150
high-residue crops 77
Hillary, Sir Edmund 173
Himalayas 174
Högbom, Arvid 2
holding tanks 70
home energy use 116
home heating 79
honey 177
Hopwood, Nick 19–20
horseshoe crabs 160
hot spots, biodiversity 110–112
House, Kurt Zenz 223
human activity. *See* anthropogenic causes/effects of greenhouse gases
Huntsman, Jon 218
Hurricane Katrina *191,* 238*c*
hurricanes 190
hydrofluorocarbons (HFCs) 20–22

I

ice age 2–4, 234*c*
icebergs 146
ice core studies 13, 235*c*–*237c*
Iceland 52
ice shelves 182–183

Inconvenient Truth, An (film) 238*c*
India 173
industrialization 173
Industrial Revolution 4, 10
industry, CO_2 and 38–39
infrared radiation 3–5, 7, 9–10
Inkley, Doug 154
Inouye, David 153
insects 106
Inslee, Jay 168
insolation 3, 5–6, 27
insurance industry 139–140
Intergovernmental Panel on Climate Change (IPCC) 236*c*
international effects of climate change 171–203
invasive species 64, 165, 196
IPCC (Intergovernmental Panel on Climate Change) 236*c*
irrigation 86, 196
Italy 177–178

J

jaguarrundi 163
Janghu Sherpa, Tashi 174
Johns, Andrew 109
Juan de Fuca Plate 52, 221

K

Kalnay, Eugenia 121
Kanter, Nechy 53
Katrina, Hurricane *191,* 238*c*
Karoly, David 180
Keeling, Charles David 12, 15, 235*c*
Keeling, Ralph 15
Keeling Curve 13–14, *14,* 235*c*
Keipper, Vince 179
Kentucky 50–51, 119
Kilimanjaro, Mount 179
Kinsinger, Anne 167
Kirshen, Paul 128
Kwinana Desalination Plant 231
Kyoto Protocol 237*c,* 238*c*
- and carbon sinks 98
- and CDM 90–91
- United States and 48–49, 92, 143, 231–232

L

lag time. *See* thermal inertia
land management
- after deforestation 109–112
- and agricultural sequestration 43
- and crop productivity 83
- and ecosystem health 94
- in southeastern U.S. 191–192

land-use change 78, 119–122
La Niña 190
latent heat 120
laws 145–146, 154–155, 166–170
leakage 43
Lieberman, Joseph 238*c*
lifestyle choices. *See* carbon footprint
lime 35
limestone 53, 55
Little Ice Age 234*c*
livestock 19, 63–66, 69–70
lobsters 157
loggerhead sea turtle 152, 164
logging 88, 93, 105, 189
Loladze, Irakli 67
low elevation coastal zone 129
Lower Rio Grande Valley National Wildlife Refuge 163
low-till farming 75
Lubber, Mindy 128–129

M

MacCracken, Michael 226
magnesite 55
magnesium 24, 53, 55
Maldives 132
manure 20, 69–73
Marchetti, C. 219
market-based incentives 90
Marra, Peter 155–157
Mauna Loa CO_2 monitoring station (Hawaii) 12–15, *13,* 235*c*
McCain, John 238*c*
McCarl, B. A. 40, 66
McKay, Simon 231
McKinsey and Company 224–225
McLeish, Derek 57
McPherson, Brian 218
medicinal plants 108. *See also* pharmaceuticals
meltwater 149
Mendelsohn, Robert 66
methane
- from anaerobic digester systems 70–71
- and carbon cycle 32
- chemistry/sources of 19–20
- and enhanced greenhouse effect 10
- forest fires and 35
- GWP of 39, 70

from livestock 19, 69–73
and sequestration 40, 55
methane hydrate 29
methanol 69
Mexican jay 153, 156
micronutrients 67
midwestern United States 192–195
Midwest Regional Carbon Sequestration Partnership 53
migration, climate change and 155–157
migratory songbirds 160, 164
mineral sequestration 53, 55
Miranti, Giuseppe 177
"missing carbon" 94
mitigation 74–79, *76, 77,* 129, 208–225
Mitloehner, Frank 69
moldboard plowing 78
monarch butterfly 149
monsoons 68
Morgan, M. Granger 52
mosquitoes 148
Mount Simon Sandstone formation 53
Muir Glacier *124*
musk ox 203

N

National Assessment 225–226, 237*c*
National Wildlife Refuge system 158–166, *161*
natural gas 38
natural greenhouse effect 10
natural refuges 141–170
Natural Resource Conservation Service, USDA 73
negative radiative forcing 16, *17*
Nelson, Frederick 206–207
Nemani, Ramakrishna 68
Nepal 174
New Direction for Energy Independence Act 168
Newell, Brent 70
New Orleans, Louisiana *191*
New York City 134–135
niche 101
Nielson, Dianne 217–218
Nisqually National Wildlife Refuge 163–164
nitrogen 67
nitrous oxide (N_2O) 10, 20, 40–41, 73, 78
Nixon, Richard M. 145
"Noah's choice" 166
Norby, Richard J. 105
Norgay, Tenzing 173
North America 185, 186. *See also* United States
North Atlantic Deep Water 182
northeastern United States 186–189
Northern Hemisphere 105
northern plains 65
Noss, R. F. 93
no-till planting *77,* 78
nutrients 109
nutritional value, of crops 66, 67

O

Obama, Barack 232–233, 239*c*
ocean acidity 210
ocean circulation 182
Ocean Conveyor Belt 182
ocean currents 33
ocean floor 49
Oceania 179–181
oceans 11, 12, 31–33, 118, 210
ocean sequestration 55–56, 213, 219–223, *220, 222*
ocean temperature 33–34
ocelot 163
oil 38, 166, 234*c,* 239*c*
oil fields, geologic sequestration and 50, 58, 212–213, 215
Ostrowski, Ken 225
ozone, ground-level 82, 190
ozone layer (stratospheric) 25–27

P

Pacala, Stephen 34
Pacific Northwest 51, 199–201
panther *162*
Parmesan, Camille 153–154
particulates 27
Payne, Roger 173–174
PCC (precipitated calcium carbonate) 56–57
Peabody Energy 50
Pea Island National Wildlife Refuge 164
peat 97
pelicans *147*
penguins 146
perfluorocarbons (PFCs) 21, 22
permafrost 36, 37, 146, 202, 205–207
Perth, Australia 231
Peterson, Bruce 181, 182
petroleum. *See* oil
Pew Center on Climate Change 61–62
PFCs. *See* perfluorocarbons
pharmaceuticals 101–103, 108–109

photosynthesis 11–12, 32–33, 36, 39, 43, 62–63
physics of greenhouse gases 18–21
phytoplankton 34–36
Piacenza, Italy 177–178
Pied flycatcher 142
pikas 150–152, *151*, 157
Pimm, Stuart 111–112
plantations 109
plant decomposition 11
Plantinga, Andrew 57–58
polar bears 143–144, *144*, 157, 226, *228*
polar regions 205
pollutants, global dimming by 27, 28
positive radiative forcing 16, *17*
prairie pothole region 146–147, 158
precipitated calcium carbonate (PCC) 56–57
precipitation
  and agriculture 62, 68, 81–82
  in Alaska 201
  deforestation and 102–104
  in Great Plains 195–196
  land use change and 121
  in midwestern U.S. 195
  in Pacific Northwest 199–201
  in southeastern U.S. 190
proactive adaptation 229
property insurance 139
prothonotary warblers 157
public lands 125–128

Q

Qinghai-Xizang railroad 207

R

radiant heating 6
radiant power 6
radiative forcing 16, *17*
Rahall, Nick 168
Rahmstorf, Stefan 182
rainfall. *See* precipitation
rain forest 102–104, 108, 110
rangeland 64, 78, 83, 85, 90
Rappahannock River Valley National Wildlife Refuge 164
Raupach, Michael 117
Ray, Chris 150, 152
reactive adaptation 229
reduced tillage 78. *See also* conservation tillage
red-winged blackbirds 157
redwoods *95*
reflection, of solar radiation 6–9, 27, 120
reforestation 91, 93, 97–98
refuges (wildlife) 158–166
regulation of agricultural industry 69–70
reindeer 203
renewable energy 75
reproduction traits of plants 67
reservoirs. *See* carbon sinks
respiration 11, 34–35
Retallack, Simon 98
Reynolds, Anna 180
Rhode Island National Wildlife Refuge Complex 164
rice cultivation 20
Rio Summit (1992) 236*c*
Roberts, Kerry 231
robins 153, 156
Robock, Alan 118
rockslides 207
Rocky Mountain National Park 150
roofs, for and urban heat mitigation 135, 136
Root, Terry 112, 142, 143, 158
Rosenbaum, Barry 151
Rudd, Kevin 51
ruminants 19, 69
runoff 195
Russia 205–206

S

Safe Climate Act of 2007 167–168
Sahel 28
San Luis National Wildlife Refuge 165
Saxton, James 168
Schimel, David 54, 68
Schneider, Steve 142–143
Schneider, Uwe 40
Schwarzenegger, Arnold 218
seafloor sediments 53, 55
sea-level rise
  2003 studies 238*c*
  and Antarctic ice sheet loss 205
  in coastal areas 123, 129–133, *130*
  cultural losses from 138–139
  in northeastern U.S. 188
  in Oceania 179–180
  in Pacific Northwest 200–201
  in southeastern U.S. 190, 192
  underestimation of 118
seasonal events, timing of 152–158
sea turtle 152

sedimentary basins 215–216
sediments 31–32
Sedum spurium 136
seeds 67, 68
selective logging 109
semiconductor industry 24
sensible heat 120
September 11, 2001, terrorist attacks 27
sequestration. *See* carbon sequestration
serpentinite 55
severe weather. *See* extreme weather events
sewage treatment 20
sheep 147
Shrestha, Surendra 174
sinks. *See* carbon sinks
Slagle, Angela 223
slash-and-burn agriculture 109
Sleipner Gas Field 215
smelting 24
snowmelt 197
sockeye salmon 156
soil
  in boreal forest 96, 97, 109
  as carbon sink 36–37, 75
  as carbon source 11, 93
  global warming's effect on 54, 61
  and N_2O 20
soil erosion 66
soil moisture 158
solar radiation 3, 5–10
songbirds 149
sooty shearwaters 156
South Africa 139
South America 184, 185
southeastern United States 58, 189–192, *191*
species conservation 57–58
Sperber, Georg 178
spruce trees 178
Statoil 215
Stephenson, Jack 224
Stern, Sir Nicholas 238*c*
Stork, N. E. 108
storm surges 192, 202
stratosphere 25
sub-Saharan Africa 86
subsistence livelihoods 202–203
subtropics 81–82
sulfur dioxide (SO_2) 24
sulfur hexafluoride (SF_6) 21, 22, 24
Sunbelt. *See* southeastern United States
surface temperatures 4
swamp gas 19
Swetham, Tom 128
Swiss Re 139–140

T

Takahashi, Taro 56, 221, 222
tax credits 91
temperate forests 109
temperature increases. *See also* climate change
  and agriculture 62, 63, 81
  in Alaska 201
  deforestation and 103
  in developing world 74
  in Great Plains 195
  and greenhouse effect 5
  land use change and 121–122
  midwestern U.S. 193
  Pacific Northwest 199–200
terrestrial biosphere 31
thermal inertia 17–18, 226
Thompson, Lonnie 179
Tibet 207
tillage methods 75. *See also* conservation tillage
timber production 105, 189
tipping point 11, 167
Tokyo, Japan 136
tourism 126–127, 198
trace gases 5, 10, 18
Trans-Alaska pipeline *206*
transportation 38, 79, 116
tree line 150
trees 135, 152. *See also* forests
Trempealeau National Wildlife Refuge 165
Trenberth, Kevin 117–118
tropical regions 47, 90
tropical vegetation 63
troposphere 25
trout 149–150, 189
tundra 150, 201, 203
Turnbull, Malcolm 230–231
Tyndall, John 2, 234*c*

U

ultraviolet B radiation 25
underground injection 42–43, 212–213. *See also* geologic sequestration
UNFCC. *See* United Nations Framework Convention on Climate Change
United Nations Climate Change Conference 239*c*

United Nations Framework Convention on Climate Change (UNFCC) 22, 48, 91
United States 186–203, *187*
  Alaska 201–203
  carbon footprint 118
  carbon sequestration 48
  carbon sinks 34, 88
  farming practices and climate change mitigation 74
  Midwest 192–195
  modeling of carbon emissions 58–59
  National Wildlife Refuge system 158–166, *161*
  Northeast 186–189
  ozone depletion 26
  Pacific Northwest 199–201
  soil freezing in 206
  Southeast 189–192, *191*
  West 196–198
United States National Assessment on the Potential Consequences of Climate Variability and Change 225–226, 237*c*
urban areas
  cement production and CO_2 35
  climate solutions for 137–138
  dangers of climate change 128–138
  northeastern U.S. 188–189
  sea-level rise and 129–133, *130*
urban heat island effect 132, 134–138, 193–194
Utah 217–219

V

vegetated roofs 135, 136
vegetation 13–14, 19, 32–33, 63, 120, 134. *See also* crop productivity
Velicogna, Isabella 183
visible light 6–9
VOCs (volatile organic compounds) 69
volatile organic compounds (VOCs) 69
volcanic rock 51
volcanoes 11, 35
Vulcan CO_2 modeling system 58–59

W

water
  absorption of visible light by 9
  and agricultural sequestration 45
  in Great Plains 196
  and greenhouse effect 10
  in midwestern U.S. 193
  in Pacific Northwest 199–200
  southeastern U.S. 192
  in western U.S. 197–198
waterfowl 146–147, 158–159
water pollution 70, 72, 162–163
water table 196
water vapor 2, 9–10, 19, 27, 235*c*
Waxman, Henry 167
Wayburn, L. A. 93
weather 61, 104. *See also* extreme weather events
Weishampel, John 152, 154
West, Tristam O. 43, 45
West Antarctic ice sheet 131, 183, 205
Western Climate Initiative 218
Western Kentucky Carbon Storage Foundation 50
western United States 196–198
wetlands destruction 131, 165, 189
wildfires 65, 94, *107,* 123, 127–128, 194, 198
wildlife 141–170, *144, 147, 151*
  climate change's effect on forests 110–112
  laws in force to help 145–146, 154–155, 166–170
  refuges at risk 158–166, *161*
  timing of seasonal events 152–158
Wildlife and Wetlands Toolkit for Teachers and Interpreters 125
Wilkins Ice Shelf 239*c*
wind energy 231, 239*c*
Wofsy, Steven C. 94, 97
World Climate Research Programme 235*c*
World War II 70

Y

yellow-bellied marmots 153
Yukon Flats National Wildlife Refuge 165–166

Z

Zhang, Tingjun 205
Zion National Park 125